T0297800

CLINICAL INFORMATICS LITERACY

CLINICAL INFORMATICS LITERACY

5000 Concepts That Every Informatician Should Know

DEAN F. SITTIG, PhD

School of Biomedical Informatics
University of Texas, Houston, TX, United States

ACADEMIC PRESS

An imprint of Elsevier

Academic Press is an imprint of Elsevier
125 London Wall, London EC2Y 5AS, United Kingdom
525 B Street, Suite 1800, San Diego, CA 92101-4495, United States
50 Hampshire Street, 5th Floor, Cambridge, MA 02139, United States
The Boulevard, Langford Lane, Kidlington, Oxford OX5 1GB, United Kingdom

Library of Congress Cataloging-in-Publication Data
A catalog record for this book is available from the Library of Congress

British Library Cataloguing-in-Publication Data
A catalogue record for this book is available from the British Library

ISBN: 978-0-12-803206-0

For information on all Academic Press publications visit our website at
https://www.elsevier.com/books-and-journals

Working together
to grow libraries in
developing countries

www.elsevier.com • www.bookaid.org

Publisher: Mica Haley
Acquisitions Editor: Rafael Teixeira
Editorial Project Manager: Mariana Kühl Leme
Production Project Manager: Chris Wortley
Designer: Victoria Pearson Esser

Typeset by TNQ Books and Journals

QUOTE FROM TS ELIOT

The vast accumulations of knowledge—or at least of information—deposited by the nineteenth century have been responsible for an equally vast ignorance. When there is so much to be known, when there are so many fields of knowledge in which the same words are used with different meanings, when everyone knows a little about a great many things, it becomes increasingly difficult for anyone to know whether he knows what he is talking about or not. And when we do not know, or when we do not know enough, we tend always to substitute emotions for thoughts.

T.S. Eliot (1888–1965). The Sacred Wood. 1921.
http://www.bartleby.com/200/sw2.html.

QUOTE FROM TS ELIOT

CONTENTS

CATEGORY DEFINITIONS

ACKNOWLEDGMENTS

The following people were instrumental in helping me identify concepts to be included in this book: Emily Campbell, JoAnn Kaalaas-Sittig, Allison B. McCoy, and Daniel G. Miller, Adam Wright. In addition, I reviewed the glossaries from the following clinical informatics textbooks: Coeira's *Guide to Health Informatics, Third Edition*, Shortliffe and Cimino's *Biomedical Informatics: Computer Applications in Health Care and Biomedicine, Third Edition*, Rowland's *A Practitioner's Guide to Health Informatics in Australia.*

ABOUT THIS BOOK

This book was modeled after Hirsch's book *Cultural Literacy: What Every American Needs to Know,* which contains over 5000 names, phrases, dates, and concepts that every American should know to consider themselves culturally literate. After reading this book, I decided that there must be at least 5000 concepts that every clinical informatician should know as well. This book represents my attempt to develop such a list.

What is Clinical Informatics?

Clinical informatics is the relatively new scientific field that focuses on the sociotechnical aspects of the use of information and information technology to study and improve the health of individuals and the organizational and technical systems that support them in that endeavor. While many have tried to prove that clinical informatics and the artifacts it creates can directly improve the health of individuals or the health of large populations of individuals, in my experience such improvements in health or health care are an indirect result of the work of clinical informaticians. Therefore, by definition, clinical informatics is a multidisciplinary field that requires widespread clinical and technical knowledge as well as the ability to work alongside expert-level clinicians, technologists, and healthcare administrators.

This book is designed to help those interested in the field of clinical informatics to understand the breadth of knowledge required to successfully participate in the design, development, implementation, use, and evaluation of the health information technology (HIT) required to transform the current complex adaptive healthcare system into a robust, reliable, efficient, and cost-effective HIT-enabled healthcare system. Such a transformation will require concerted effort on the part of many individuals who each bring unique knowledge, skills, and experience to bear on the myriad problems that must be identified, defined, explored, and overcome.

A Sociotechnical Approach to Clinical Informatics

As previously stated, clinical informatics is a sociotechnical field that is well-described by an eight-dimension sociotechnical model that Hardeep Singh, MD, MPH, and I developed to help clinicians, technologists, and researchers understand the various sociotechnical aspects of the field and their complex interactions. The following sections (adapted from Sittig DF, Singh H. A new sociotechnical model for studying health information technology in complex adaptive healthcare systems. Quality & Safety in Health Care. 2010 Oct; 19 Suppl 3:i68–74. http://dx.doi.org/10.1136/qshc.2010.042085) describe each of these eight dimensions, first in terms of what is meant by each dimension and second, why each dimension is so important to understanding the complexity of the field. Included within each dimension are a few examples of the categories of concepts related to that dimension.

Hardware and Software Computing Infrastructure

This dimension of the model focuses solely on the hardware and software required to run the clinical informatics applications. The most visible part of this dimension is the computer, including the monitor, printer, and other data display devices along with the keyboard, mouse, and other data entry devices used to access clinical applications and medical or imaging devices. This dimension also includes the centralized (network-attached) data storage devices and all of the networking equipment required to allow applications or devices to retrieve and store patient data. Also included in this dimension is software at both the operating system and application levels. Finally, this dimension of the model subsumes all the machines, devices, and software required to keep the computing infrastructure functioning such as the high-capacity air conditioning system, the batteries that form the uninterruptible power supply (UPS) that provides short-term electrical power in the event of an electrical failure, and the diesel-powered backup generators that supply power during longer outages. In short, this dimension is purely technical; it is only composed of the physical devices and the software required keeping these devices running.

One of the key aspects of this dimension is that, for the most part, the end-users are not aware that most of this infrastructure exists until it fails. Therefore, everyone working in the field of clinical informatics must have at least a passing knowledge and understanding of the design, development, implementation, use, and monitoring of the equipment and methods used to keep the computer applications running. Likewise, since the entire computing industry continues to move forward with astonishing speed, clinical informaticians need to be aware of the latest developments and improvements in the hardware and software they are relying on. Often, what was virtually impossible several years ago, due to inadequate processing power (e.g., real-time monitoring of all in-patients to identify potential cases of sepsis via a remote-hosted service), data storage capacity (i.e., real-time access to all patients complete history of imaging procedures, clinical notes and reports), or networking bandwidth (i.e., real-time broadcast of telemedicine-enabled clinical procedures around the world) can now be accomplished relatively, easily, and cheaply using commercially available off-the-shelf hardware and software. Therefore, readers interested in exploring the hardware aspects of this dimension more fully could turn to the computer hardware, architecture, or networking categories. Those interested more in the software side of this dimension could review the computational algorithm, application development, data structure, and data analysis categories.

Clinical Content

This dimension includes everything on the data–information–knowledge continuum that is stored in the computing system. For example, data such as structured and unstructured textual or numeric data and images that are either captured directly from imaging devices or scanned from paper-based sources; information such as online clinical reference resources that are available to clinicians at the point of care to help them remember or learn important clinical concepts; and knowledge such as clinical algorithms used to generate real-time clinical alerts or disease-specific, clinical documentation templates. Various clinical content elements can be used to configure certain software applications to meet clinical or administrative requirements. Examples include controlled vocabulary items that are selected from a list while ordering a medication or a diagnostic test, and the logic required to generate an alert for certain types of medication interactions. These elements may also describe certain clinical aspects of the patients' condition (e.g., laboratory test results, discharge summaries, or radiographic images). Other clinical content, such as demographic data and patient location, can be used to manage administrative aspects of a patient's care. These data can be entered (or created), read, modified, or deleted by authorized users and stored either on the local computer or on a network-attached device. Certain elements of the clinical content, such as those which inform clinical decision support (CDS) interventions, must be carefully managed and updated on a regular basis.

As the field of clinical informatics progresses, the importance of having access to accurate, up-to-date, clinical content cannot be overemphasized. The translation of this data, information, and knowledge into computer interpretable and usable forms is one of the main challenges of the field of clinical informatics. Creation, maintenance, and utilization of this computer-based clinical content requires (1) knowledge of the way the computer algorithms and systems work, as well as, (2) a good understanding of the basic physiological, pathological, and anatomical information and knowledge required to care for patients, combined with an understanding of how clinical and administrative work is accomplished within the healthcare delivery system. These requirements help explain why approximately 30% of the concepts included in this book are from the basic biological and clinical sciences. It also explains why so many of the most successful clinical informaticians come from a clinical background (i.e., physicians, nurses, laboratory technicians, pharmacists, etc.). Finally, it means that those clinical informaticians from more technical backgrounds (e.g., computer science, engineering, statistics, information management, etc.) must learn as much about clinical science, medicine, and how the healthcare system works, as possible to be conversant with the clinicians and administrators that will be using the clinically focused systems that are developed. Therefore, those working on developing or using clinical content could review the categories describing: terminology, Unified Medical Language System vocabulary, clinical decision support, body system, disease, and clinical specialty.

Human–Computer Interface

An interface enables unrelated entities, such as humans, to interact with the computer system and includes aspects of the computer system that users can see, touch, or hear. The hardware and software "operationalize" the user interface; provided these are functioning as designed, any problems with using the system are likely due to human–computer interaction (HCI) issues. The HCI is guided by a user interaction model created by the software designer and developer and hopefully agreed to by the user community. During early pilot testing of the application in the target clinical environment, both the user's workflow and the interface are likely to need revisions. This process of iterative refinement, wherein both the user and user interface may need to change, should culminate in a HCI model that matches the user's modified clinical workflow while enabling the computer to manage the required data safely and securely. For example, if a clinician wants to change the dose of a medication, the software requires the clinician to discontinue the old order and enter a new one, but the user interface should hide this complexity. This dimension also includes the ergonomic aspects of the interface. If users are forced to use a computer mouse while standing, they may have difficulty controlling the pointer on the screen because they are moving the mouse using the large muscles of their shoulder rather than the smaller muscles in the forearm. Finally, the lack of a feature or function within the interface represents a potential problem with the interface, the clinical content that provides the selection options for the users, or with the software or hardware that implements the interface.

The HCI is one of the key dimensions of the sociotechnical model in that it is the main site at which the users, or social component of the model, interact with the technical or hardware, software, and clinical content. While many users complain about the user interface, the root of the problem may reside in another dimension of the sociotechnical model altogether. Working to understand how the various dimensions of the sociotechnical model interact, often through the user interface is another key challenge for clinical informaticians. Therefore, informaticians need to have a firm grasp of the concepts involved in designing, creating, configuring, maintaining, and evaluating the human–computer user interface. The following categories may prove useful: HCI and computer application.

People

This dimension represents the humans (e.g., software developers, system configuration and training personnel, clinicians, and patients) involved in all aspects of the design, development, implementation, and use of HIT. It also includes the ways that systems help users think and make them feel. Although user training is clearly an important component of the user portion of the model, it may not by itself overcome all user-related problems. Many "user" problems actually result from poor system design or errors in system development or configuration. In addition to the users of these systems, this dimension includes the people who design, develop, implement, and evaluate these systems. For instance, these people must have the proper knowledge, skills, and training required to develop applications that are safe, effective, and easy to use. This is the first aspect of the model that is purely on the social end of the sociotechnical spectrum.

In most cases, users will be clinicians or employees of the health system. However, with recent advances in patient-centered care and development of personal health record systems and "home monitoring" devices, patients are increasingly becoming important users of HIT. Patients and/or their caregivers may not possess the knowledge or skills to manage new health information technologies, and this is of specific concern as more care shifts to the patient's home.

The people dimension is critical for the successful application of clinical informatics' interventions within the modern day electronic health record (EHR)–enabled healthcare system. Failure to understand the roles, culture, knowledge, training, and emotional states of the people involved in building and using these complex systems will surely lead to failure of the project. Therefore, it is vitally important that clinical informaticians learn enough of the vocabulary of both the information technology professionals (i.e., technical terms in this book) responsible for building, implementing, and maintaining these systems as well as the clinical professionals (i.e., biomedical terms in this book) that will be using them. Specific categories related to this dimension include: people and organization.

Workflow and Communication

This is the first portion of the model that acknowledges that people often need to work cohesively with others in the healthcare system to accomplish patient care. This collaboration requires significant two-way communication. The workflow dimension accounts for the steps needed to ensure that each patient receives the care they need at the time they need it. Often, the clinical information system does not initially match the actual "clinical" workflow. In this case, either the workflow must be modified to adapt to the HIT, or the HIT system must change to match the various workflows identified.

This dimension highlights the importance of studying both the ways and means that humans use to communicate with each other as well as the way they carry out their work. The goal when developing new health information technology applications is to improve or facilitate communication between the key members of the healthcare system. Likewise, these new applications should make the existing workflows more efficient, safe, and effective. Failure to understand the current and future workflows of clinicians often results in failure to use the new technology as anticipated, which often show up as work-arounds. Key categories related to this dimension include: workflow, communication, and system implementation.

Internal Organizational Policies, Procedures, and Culture

The organization's internal structures, policies, and procedures affect every other dimension in our model. For example, the organization's leadership allocates the capital budgets that enable the purchase of hardware and software, and internal policies influence whether and how offsite data backups are accomplished. The organizational leaders and committees who write and implement IT policies and procedures are responsible for overseeing all aspects of HIT system procurement, implementation, use, monitoring, and evaluation. A key aspect of any HIT project is to ensure that the software accurately represents and enforces, if applicable, organizational policies and procedures. Likewise, it is also necessary to ensure that the actual clinical workflow involved with operating these systems is consistent with existing policies and procedures. Finally, internal rules and regulations are often created in response to the external rules and regulations that form the basis of the next dimension of the model.

This dimension highlights the importance for all clinical informaticians to have at least a basic understanding of how the healthcare delivery system functions. For example, they need to know the different types of healthcare facilities that exist, how they are organized, how decisions within them are made, and which key stakeholders in the organization must be consulted before any decision that might affect the health information technology that is in place or being considered for widespread implementation is made. Therefore, readers interested in this dimension should review the management, medical billing, and medical facility categories.

External Rules, Regulations, and Pressures

This dimension accounts for the external forces that facilitate or place constraints on the design, development, implementation, use, and evaluation of HIT in the clinical setting. For example, the recent passage of the American Recovery and Reinvestment Act (ARRA) of 2009, which includes the Health Information Technology for Economic and Clinical Health (HITECH) Act, made available over $35 billion dollars for healthcare practitioners who became "meaningful users" of health IT. Thus, ARRA introduced the single largest financial incentive ever to facilitate EHR implementation. Meanwhile, a host of federal, state, and local regulations regulate the use of HIT. Examples include the 1996 Health Insurance Portability and Accountability Act (HIPAA), recent changes to the Stark Laws, and restrictions on secondary use of clinical data. Finally, there are three recent national developments that have the potential to affect the entire healthcare delivery system in the context of HIT. These include: (1) the initiative to develop the data and information exchange capacity to create a national health information network; (2) the open notes initiative to enable patients to access copies of the clinical data via personal health records; and (3) government incentives to define and address clinical and IT workforce shortages.

Understanding the forces external to the actual healthcare organization, that is, implementing various health information technology interventions is critical to understanding why certain work is performed in a particular manner, or why certain work-flows are not permitted. Often these rules and regulations place enormous pressure and constraints on healthcare organizations with seemingly little input from the healthcare workers them-selves. Then these externally mandated constraints are in turn enforced by various health information technology tools and procedures often resulting in extreme user frustration. Failure to understand the root cause of these frustrations can lead infor-maticians to work on solutions that are not useful or even against the law. Readers curious about this dimension should review con-cepts in the categories of professional organization, government funding, government organization, health finance and insurance.

System Measurement and Monitoring

This dimension has largely been unaccounted for in previous models. We posit that the effects of HIT must be measured and monitored on a regular basis. An effective system measurement and monitoring program must address four key issues related to HIT features and functions. First is the issue of availability—the extent to which features and functions are available and ready for use. Measures of system availability include response times and percent uptime of the system. A second measurement objective is to determine how the various features and functions are being used by clinicians. For instance, one such measure is the rate at which clinicians override CDS warnings and alerts. Third, the effectiveness of the system on healthcare delivery and patient health should be monitored to ensure that anticipated outcomes are achieved. For example, the mean hemoglobin A1C (HbA1c) value for all diabetic patients in a practice may be measured before and after implementation of a system with advanced CDS features. Finally, in addition to measuring the expected outcomes of HIT implementation, it is also vital to identify and document unintended consequences that manifest themselves follow-ing use of these systems. For instance, it may be worthwhile to track practitioner efficiency before and after implementation of a new clinical charting application. In addition to measuring the use and effectiveness of HIT at the local level, we must develop the methods to measure and monitor these systems and assess the quality of care resulting from their use on a state, regional, or even national level.

As the percentage of the nation's gross domestic product that is spent on health care continues to increase, there is going to be even more scrutiny of the costs and effectiveness of the health care that is being delivered. The rapid adoption of health information technology is no exception. Informaticians must be ready to demonstrate that the systems they are putting into place are having a significant positive impact on the health of individuals in our country and the costs associated with keeping them healthy. To do this, informaticians need to be familiar with concepts in the categories of statistics, study design, and theory at the very least.

How to Use This Book...

This book is divided up into 80 categories. Each category contains from 10 to 200 concepts. Each category is preceded by a short description of the concepts included in that section.

This book has several different uses. The first is for self-assessment and improvement. In this mode, readers review the concepts within a particular category. I would recommend that readers with a technical background begin with the more biomedically oriented concepts. This will help ensure that they are capable of communicating more clearly with their clinical colleagues. They may also find it interesting to review some of the technical categories since many of the concepts are evolving rapidly and many new ones are being developed every day. Similarly, I would recommend that readers with a biomedical background begin with the more technically oriented concepts. As you read through the concepts, whenever you come across one that you are not familiar with, then you should go to Google and look it up. If you type *define: "depth-first search"*, for example, you will be shown a set of pages that attempt to define this concept. You should then try to use this concept in a conversation in the next few days.

The second type of user is one who is attempting to assess the breadth or depth of clinical informatics knowledge of a particular individual. This person may be an informatics student or a job candidate. In either case there are several options to consider. The first is to ask the person to define several terms that are either randomly or purposively chosen from one or more categories. This might take the form of randomly selecting 20 concepts from the book and asking a person to provide a short definition of each concept from memory. Alternatively, one could give the person the same assignment as a take-home assignment and allow them to use the internet to help them define each concept. I would expect an advanced informatics student to be able to define approximately 80% of the concepts from memory.

An alternative would be to randomly select 5–10 categories and ask the student to list five example concepts that might be included in each category. A related assessment method would be to give the student a list of 50 randomly identified concepts along with the list of all categories from the book and ask the student to put each concept into the correct category.

Similarly, one could ask a student to compare and contrast two or more concepts within the same category, for example, the programming languages "Java" and "Fortran".

Finally, one could utilize the book as a study guide. For example, one could randomly select a page and begin reviewing concepts one after the other until you come across a concept that is unfamiliar to you. Then by all means look up all unknown concepts. If one were really ambitious, then one could make flash cards using 3"× 5" index cards or one of the new online resources such as Anki (http://ankisrs.net/).

CATEGORY
DEFINITIONS

Academic Degree

A qualification, usually determined by the successful completion of a prescribed course of study in higher education that often includes the passing of a comprehensive examination. Academic degrees are normally awarded by a college, university, or any number of professional schools such as medical, nursing, dental, osteopathic, pharmacy, and public health, for example. These institutions commonly offer degrees at various levels, typically including associate (most often a 2-year course of study is required), bachelor (4-year course of study), master (1–2 year course of study after the bachelor'), and doctorate (3–7 year course of study after bachelor's or master's degree).

Bachelor of Arts (BA)
Bachelor of Medicine (BM or MB)
Bachelor of Medicine, Bachelor of Surgery (MBBS or MBChB)
Bachelor of Science (BS)
Doctor of Education (EdD)
Doctor of Jurisprudence (JD)
Doctor of Medicine (MD)
Doctor of Naturopathy (ND)
Doctor of Nursing Practice (DNP)
Doctor of Nursing Science (DNS)
Doctor of Optometry (OD)
Doctor of Osteopathic Medicine (DO)
Doctor of Pharmacy (PharmD)
Doctor of Philosophy (PhD)
Doctor of Podiatric Medicine (DPM)
Doctor of Public Health (DPH)
Doctor of Science (DSc)
Master of Arts (MA)
Master of Business Administration (MBA)
Master of Dental Science (MScD)
Master of Health Administration (MHA)
Master of Nursing (MN)
Master of Public Health (MPH)
Master of Science (MS or MSc)
Master of Science in Dentistry (MSD)
Master of Science in Nursing (MSN or MScN)
Master of Science in Pharmacy (MPh or Mpharm or MScPh)
Master of Science in Social Work (MSW)
Master of Surgery (MS)
Medical Doctorate (MD)

Clinical Informatics Literacy. http://dx.doi.org/10.1016/B978-0-12-803206-0.00001-8

Anatomy

The branch of biomedical science concerned with the bodily structure of humans, animals, and other living organisms. Anatomy is often studied through dissection and separation of individual parts of the body. For an in-depth overview of human anatomy, see: http://www.innerbody.com/.

Afferent
Alveolus
Amygdala
Aneurysm
Anterior (ventral)
Anulus
Aorta
Arteries
Artery
Atrium
Axon
Biceps brachii
Blood
Bone marrow
Both eyes (OU)
Cardiac region
Cartilage
Caudal
Central
Cephalic
Cerebral
Cerebrovascular
Cervix
Coronal plane (frontal)
Cortex
Cranial nerves
Cranial region
Deltoid
Dendrite
Diaphragm
Dissect
Distal
Dorsal
Endosteum
Esophagus
External (superficial)
Gastrointestinal (GI) tract
Gluteus maximus

Hair
Heart
Hormones
Humeral
Inferior (caudad)
Innervate
Internal
Interstitial
Intestines
Intraperitoneal
Kidneys
Lateral
Latissimus dorsi
Left eye (OS)
Liver
Lung
Lymph node
Macroscopic
Medial
Membrane
Mouth (Os)
Muscles
Nails
Nerve
Pectoralis major
Periosteum
Peripheral
Placenta cord membranes
Plasma
Posterior (dorsal)
Proximal
Pylorus
Quadriceps femoris
Red blood cell (RBC)
Renal
Right eye (OD)
Right lower arm (RLA)
Right lower quadrant (RLQ)
Right upper quadrant (RUQ)
Sagittal plane
Septum
Serum
Sigmoid colon
Sketch
Skin

Stomach
Striated
Superior (cephalad)
Sweat
Syncytium
Trachea
Transverse plane (axial or cross section)
Triceps brachii
Unilateral
Veins
Vena cava
Ventral
Ventricle
Visceral
Vivisection
White blood cell (WBC)

Application Development

A field of study that includes the set of processes, procedures, and practices of developing software applications. Depending on the size, complexity, and criticality of the application to be developed, the process may involve the use of one or more programming languages, application development frameworks, testing methodologies, and one or more teams of software developers.

Agile software development
Capability Maturity Model (CMM)
Data modeling
Design effect
JavaScript Object Notation (JSOM)
Joint applications design (JAD)
Logical data model (LDM)
Logical schema
Productivity
Rapid application development (RAD)
Rapid prototyping
Requirements analysis
Software Engineering Institute Capability Maturity Model (SEI-CMM)
Software quality assurance (SQA)
Software risk analysis
Spiral software development
Subject-matter expert (SME)
Waterfall method

Artificial Intelligence

A subfield of computer science that focuses on the design, development, use and evaluation of computer-based systems, applications, and algorithms that mimic cognitive processes usually associated with human intelligence. The origins of the field of clinical informatics were in the field of artificial intelligence as researchers attempted to create computer systems that could diagnosis patients' medical conditions. In the late 1980s, after several large-scale, highly visible AI projects failed to meet overly optimistic expectations, federal and commercial funding for new AI project rapidly dried up. This lead to the so-called AI winter. During this period, many AI researchers turned to building much less ambitious "expert systems" that proved very successful. These expert systems were further simplified to what became basic clinical decision support functionality that was widely implemented directly in electronic health records to perform simple drug–drug interaction checks or generate health maintenance reminders. In the early 2000s, with advent of the "big-data" revolution, several AI-type diagnostic decision support systems began to reappear.

Abduction
All source intelligence
Authoring system
Background question
Case-based reasoning (CBR)
Causal reasoning
Chance node
Conceptual knowledge
Connectionism
Consulting model
Consulting system
Critiquing model
Deduction
Evoking strength
Explicit
Facts
Factual knowledge
First principles, reasoning from
Foreground question
Frequency weight
HELP sector
Heuristic
Hypothetico-deductive approach
Immersive simulated environment
Implicit

Import number
Induction
Inference
Influence diagram
Integrative model
Knowledge-based system
Logical positivism
Model-based reasoning
Modus ponens (Latin for "mode that affirms")
Modus tollens (Latin for "mode that denies")
Overfitting
Problem solver
Problem space
Problem-solving method (PSM)
Prognostic scoring system
Proposition
Qualitative reasoning
Reasoning
Reasoning about time
Reminder systems
Representation
Rule interpreter
Secondary knowledge-based information
Situation action rules
Skeletal plans
Standard gamble
State diagram
Symbol
Treatment threshold probability
Truth maintenance

Body System

The human body's key systems are composed of collections of cells, tissues, and organs that work together for a common purpose. Each system performs a key role in helping the body to work effectively.

Cardiovascular system
Central nervous system (CNS)
Circulatory system
Digestive system
Endocrine system
Excretory system
Exocrine system
Immune system
Integumentary system
Lymphatic system
Muscular system
Nervous system
Olfactory system
Renal system
Reproductive system
Respiratory system
Skeletal system
Urinary system

Bone

Hard, dense, rigid, yet lightweight and strong, whitish, active, connective tissue that makes up the human skeleton, supports and protects the organs of the body, produces red and white blood cells, stores minerals, and enables mobility. Bones come in a wide variety of sizes and shapes and have a complex three-dimensional internal and external structure. The mineralized matrix of bone tissue has an organic component, mainly collagen, and an inorganic component of bone mineral made up of various salts. In the adult human there are 206 separate bones. The largest bone in the human body is the thighbone (femur) and the smallest is the stapes in the middle ear.

Carpals
Cervical ribs
Cervical vertebrae
Clavicle
Coccyx
Costae (ribs)
Cranial bones
Cranium
Femur
Fibula
Frontal bone
Humerus
Lacrimal bone
Lumbar vertebrae
Mandible (lower jaw)
Maxillae (upper jaw)
Metacarpals
Metatarsals
Nasal bones
Occipital bone
Palatine bone
Parietal bones
Patella (knee cap)
Pelvis
Phalanges
Radius
Sacrum
Scapula
Stapes
Sternum
Temporal bones
Thoracic vertebrae

Tibia (shin)
Ulna
Vertebrae
Vertebral column
Zygomatic bone

Chemistry

The branch of science that deals with the identification of the substances of which matter is composed. Chemists also investigate the properties of these substances and the ways in which they interact, combine, and change. Finally chemists study the use of these processes to form new substances. To find specific information about various facets of the field of chemistry, see: http://www.chemistryguide.org/.

0°C (freezing point of water)
32°F (freezing point of water)
100°C (boiling point of water)
212°F (boiling point of water)
Acid
Activation energy
Anion
Anode
Aqueous
Atmospheric air
Avogadro's number
Base
Buffer solution
Capacitance
Cation
Cofactor
Concentration
Conductance
Conductivity
Countercurrent
Diffusion coefficient
Electroneutrality
Electrolyte
Filter (for physical material)
Fluorescent
Flux
Half-life
Homogeneous
Isolated
Isotonic
Lyse
Medium
Modulator
Molality
Molarity
Noxious

Osmolarity
Partial pressure in a gas mixture
Permeability
pH
Potentiation
Preparation
Rate constant
Relative humidity
Resistance
Sink
Tonicity
Trace
Turbid
Turbulence
Vapor pressure
Wavelength

Clinical Decision Making

The cognitive process is used by clinicians to decide what is wrong with the patient, what should be done to remedy or alleviate the patient's problem, and when these interventions or procedures should be performed. Often there are many elements of uncertainty in the decision-making process. Therefore, clinicians must assess the probability that a particular patient is (or is not) suffering from a particular illness along with the potential harm that could occur if he or she is wrong. Wrong can be defined as either the patient has a treatable illness and he or she does not recognize it, or the patient is treated for a particular illness that he or she does not have.

Anchoring bias
Ascertainment bias
Assessment bias
Availability bias (or heuristic)
Bayesian approach
Bias
Clinical guideline
Clinical judgment
Cognitive bias
Cognitive heuristics
Concordant (test results)
Confirmation bias
Context
Decision analysis
Decision node
Decision tree
Expected utility
Expected value decision making
Indifference probability
Knowledge
Life expectancy
Pathognomonic
Prophylactic
Protocol (care plan)
Quality-adjusted life years (QALYs)
Recency bias
Referral bias
Reflective thinking
Risk attitude
Risk neutral
Shared decision-making
Summative decision

Test interpretation bias
Test referral bias
Utility
Withholding/withdrawing treatment

Clinical Decision Support

Clinical decision support (CDS) is a category of concepts and methods designed to provide patient-specific clinical information to a healthcare provider at the point of care. The goal of CDS is to improve the quality, safety, and reliability of the care provided while at the same time reducing its cost. CDS can take the form of many different types of interventions within an electronic health record. For example, order sets, condition-specific clinical displays, access to reference information, and clinical alerts are all types of CDS that have been designed and developed since the early 1960s. In addition, in the early days of the field of clinical informatics there was a concerted effort to develop diagnostic decision-support systems that would help clinicians create a differential diagnosis and eventually identify the patient's diagnosis. Although the systems were shown to be nearly as effective as expert clinicians, they fell out of favor in the late 1980s and early 1990s. More recently several companies have developed new products using similar techniques, and these applications are slowly gaining a following and have potential to offer high-quality advanced CDS regarding diagnoses to clinicians.

Action item
Action palette
Admission order sets
Alert acceptance rate
Alert fatigue
Alert message
Alert notification
Alert override rate
Alert salience
Alert trigger
Alerts
Antecedent
Antibiotic ordering support
ASBRU—clinical guideline representation language
Automated decision support
Automatic order termination
Backward chaining
Beer's criteria
Black box warnings
Care reminders
Careflow
Clinical content
Clinical content providers
Clinical decision support system (CDSS)

Clinical information online resources
Clinical pathway guideline (CPG)
Clinical Practice Guideline–Reference Architecture (CPG-RA)
Clinical prediction rule
Cognitive artifacts
Computer interpretation
Computer-interpretable guideline (CIG)
Condition-specific order sets
Condition-specific treatment protocol
Consequent
Consultation systems
Context-sensitive information retrieval
Context-sensitive user interface
Cookbook medicine
Critical lab value checking
Critiquing systems
Decision support opportunity map
Declarative knowledge
Default doses/pick lists
Departmental order sets
Description logic
Diagnostic support
Digital electronic Guideline Library framework (DeGeL)
Disease-specific order sets
Documentation aids
Drug/allergy interaction checking
Drug/condition interaction checking
Drug/drug interaction checking
Duplicate order checking
e-Mycin
EON
Evidence grading
Evoking criteria
Expression language
Five rights of clinical decision support
Formalism
Formulary checking
Forward chaining
Framingham equation
Free-text order parsing
Guideline
Guideline Elements Model (GEM)
Guideline Expression Language (GELLO)
Guideline Markup Tool (GMT)
Hard stop

High-risk state monitoring
IBM's Watson
Implication
Indication-based ordering
Interpret
Interpretation systems
Interruptive alert
Intrusive alert
IV/PO conversion
Knowledge acquisition
Knowledge base
Knowledge discovery
Knowledge engineering (KE)
Knowledge management (KM)
Knowledge modeling
Knowledge representation
Laboratory test interpretation
Look-alike/sound-alike medication warnings
Maximum daily dose checking
Maximum lifetime dose checking
Medical logic module (MLM)
Medication/laboratory test cost display
Medication dictionary
Medication dose adjustment
Medication order sentences
MediConsult
Modal alert
Monitoring systems
Nomogram
Noninterruptive alert
Nonintrusive alert
Nonmedication order sentences
Notify me when
Nutrition ordering tools
Order approvals
Order routing
Order sets
Patient-specific relevant data displays
Personal order sets
Plan of care alerts
Polypharmacy alerts
Preventive care reminders
Problem list management
Procedural knowledge
Procedure-specific order sets

Prognostic tools
Quality metric
Question prototypes
Radiology ordering support
Reference links
Registry functions
Representation of time
Risk assessment tools
Risk calculator
Service-specific order sets
Single dose range checking
Standards-Based Sharable Active Guideline Environment (SAGE)
Standing orders
Subsequent or corollary orders
Syndromic surveillance
Synthesize
Systematic review
Tacit knowledge
Tallman Lettering
Task-network model (TNM)
Ticklers
Transfer order set
Transfusion support
Treatment planning
Triage tools
Trigger event
Virtual medical record (vMR)
Weight-based dosing

Clinical Disorder

A functional abnormality or disturbance in one or more parts of the human body. Clinical disorders can be categorized into mental disorders, physical disorders, genetic disorders, emotional and behavioral disorders, and functional disorders. The term disorder is often considered more value-neutral and less stigmatizing than the terms disease or illness, and therefore is often the preferred terminology. In mental health, the term mental disorder is used as a way of acknowledging the complex interaction of biological, social, and psychological factors in psychiatric conditions.

Abdominal and pelvic pain
Abdominal aortic aneurysm (AAA)
Abnormal uterine bleeding
Above the knee amputation (AKA)
Acute kidney injury (AKI)
Acute myocardial infarction (AMI)
Alcohol abuse (EtOH)
Alzheimer disease
Anemia
Anxiety
Aortic aneurysm
Aortic stenosis (AS)
Arteriosclerosis
Arthralgias
Atelectasis
Atherosclerosis
Atrial fibrillation (Afib)
Atrial septal defect (ASD)
Attention deficit hyperactivity disorder (ADHD)
Autism spectrum disorder (ASD)
Back pain
Below the knee amputation (BKA)
Benign neoplasms
Blind
Bone pain
Cardiovascular disease (CVD)
Cervical cancer
Chest pain
Chronic condition
Chronic disease
Chronic illness
Chronic kidney disease (CKD)
Chronic obstructive pulmonary disease (COPD)
Chronically ill

Cognitive impairment
Coma
Complicated pregnancy
Congenital anomalies
Congestive heart failure (CHF)
Constriction
Coronary artery disease (CAD)
Cough
Crying
Deafness
Deep vein thrombosis (DVT)
Delirium
Delirium tremens (DTs)
Dementia
Dependence
Depression
Developmental disability (DD)
Diabetes mellitus (DM)
Diabetic ketoacidosis (DKA)
Diarrhea
Dilation
Disability
Dysphagia
Dyspnea
Dysuria
Edema
Embolism
Embolus
End-stage renal disease (ESRD)
Erectile dysfunction (ED)
Etiology
Extremity pain
Facial flushing
Facial pain
Fatigue
Fever
Fixation
Flank pain
Frustration
Functionally disabled
Funny Looking Kid (FLK)
Gallbladder disorders
Genital skin lesion
Genital ulcer
Handicapped

Hard of Hearing (HOH)
Headache
Hearing loss
Heart failure (HF)
Hematuria
Hernia
Homebound
Homicide
Hydrops fetalis
Hypertension (HTN)
Hypotension, shock
Impairment
Indication infarct
Intrauterine hypoxia
Ischemia
Labile
Labor/Delivery complications
Learning disability (LD)
Leg pain
Lesion
Lethargy
Limp
Low back pain (LBP)
Lymphadenopathy
Malaise
Malignant
Malignant neoplasms
Memory loss
Mental health
Mental illness/impairment
Mentally retarded/developmentally disabled (MR/DD)
Minimally conscious state
Mitral regurgitation (MR)
Morbid
Muscle cramps
Myalgias
Myocardial infarction (MI)
Nausea
Neonatal hemorrhage
Numbness
Nutritional deficiencies
Obsessive compulsive disorder (OCD)
Occlusion
Oppositional defiant disorder (ODD)
Otalgia

Parkinson disease (PD)
Patent ductus arteriosus (PDA)
Patent foramen ovale (PFO)
Perinatal period
Permanent vegetative state (PVS)
Petechiae
Postpartum depression (PPD)
Pregnancy
Premature atrial contractions (PACs)
Proteinuria
Pruritus
Pulmonary embolism (PE)
Pulmonary hemorrhage
Rash, generalized
Red eye
Scrotal pain
Seizure
Senility
Sensory loss
Seriously emotionally disturbed
Short gestation
Shortness of breath (SOB)
Shoulder pain
Sinus tachycardia
ST elevation myocardial infarction (STEMI)
Suicide
Syncope
Tachypnea
Tinnitus
Torticollis
Transient
Transient ischemic attack (TIA)
Traumatic brain injury (TBI)
Tremor
Tumor
Turgid
Twitch
Vasoconstriction
Venous thromboembolism (VTE)
Ventricular septal defect (VSD)
Vomiting
Weakness
Weight loss

Clinical Procedure

A clinical procedure is a physical process intended to identify a problem or achieve a result in the care of patients with health problems. Clinical procedures can be used for various reasons including: identifying, measuring, diagnosing, treating, restoring structure or function of a specific patient symptom, condition, or specific physiological parameter.

Acupuncture
Advanced cardiac life support (ACLS)
Advanced life support (ALS)
Anesthesia
Angiogram (Angio)
Angiography
Animal-assisted therapy
Antivenom
Aortography
Apheresis
Arterial blood gas (ABG)
Arterial catheter (line)
Arterial pressure
Auscultation
Basic life support (BLS)
Blood test
Cancer immunotherapy
Cancer vaccine
Cardiac stress test
Cardioconversion
Cardiopulmonary resuscitation (CPR)
Cell therapy
Central venous catheter (line)
Central venous pressure (CVP)
Cerebral angiography
Chelation therapy
Chemotherapy
Cognitive behavioral therapy (CBT)
Cold compression therapy
Combination therapy
Computer-based monitoring
Coronary angiography
Coronary arteriography
Craniosacral therapy
Cytoluminescent therapy
Diagnostic bronchoscopy
Dislocation procedure

Drug therapy
Electrocardiography
Electroconvulsive therapy
Electrocorticography
Electroencephalography
Electromyography (EMG)
Electroneuronography
Electronystagmography
Electrooculography
Electrophoresis
Electroretinography
Electrotherapy
Endoluminal capsule monitoring
Enzyme replacement therapy
Epidural (extradural) block
Esophageal motility study
Evoked potential
Extracorporeal carbon dioxide removal (ECCO2R)
Extracorporeal membrane oxygenation (ECMO)
Facial rejuvenation
Fluid replacement therapy
Fluoride therapy
Fracture procedure
General anesthesia
Heat therapy
Hemodialysis
Hemofiltration
History and physical (H&P)
Hormonal therapy
Hormone replacement therapy
Hydrotherapy
Hyperbaric oxygen therapy
Immunization
Immunosuppressive therapy
In vitro fertilization (IVF)
Infusion
Inhalation therapy
Insulin potentiation therapy
Insulin shock therapy
Intramuscular (IM)
Intravenous therapy
Intubation
Invasive
Laboratory tests
Laser therapy

Life-sustaining treatment
Lithotomy
Lithotripsy
Lithotriptor
Local anesthesia
Low-dose chemotherapy
Lymphangiography
Magnetic resonance angiogram (MRA)
Magnetic therapy
Magnetoencephalography
Mechanical ventilation
Medical inspection (body features)
Monoclonal antibody therapy
Nebulization
Negative pressure wound therapy
Nicotine replacement therapy
Noninvasive
Noninvasive monitoring technique
Ophthalmoscopy
Opiate replacement therapy
Oral rehydration therapy
Otoscopy
Oxygen therapy
Palliative care
Palpation
Particle therapy
Patient monitoring
Percussion (medicine)
Perfuse
Phage therapy
Photodynamic therapy
Phototherapy
Physical exam (Px)
Physiotherapy
Plasmapheresis
Point-of-care testing
Politzerization
Posturography
Precordial thump
Prophylactic treatment
Proton therapy
Psychotherapy
Pulmonary angiography
Radiation therapy
Radiation therapy planning

Radiography
Regional anesthesia
Respiratory therapy (RT)
Rule out (RO)
Scintillography
Shock therapy
Speech therapy
Spinal anesthesia (subarachnoid block)
Stem cell treatments
Stool test
Subclavian catheter (line)
Subcutaneous (Sub-Q)
Symptomatic treatment
Targeted therapy
Thermography
Thrombosis prophylaxis
Topical anesthesia (surface)
Tracheal intubation
Transcutaneous electrical nerve stimulation (TENS)
Treatment (tx)
Universal precautions
Unsealed source radiotherapy
Vaccination
Vaginal birth after cesarean (VBAC)
Ventriculography
Virtual reality therapy
Vision therapy

Clinical Role

In a healthcare organization there are many different jobs that need to be done. Clinicians with different training and experience do these jobs by fulfilling a "role." These clinical jobs almost always involve contact with patients. For the most part, they usually require formal study and training after you have finished high school, college, and often medical, nursing, or pharmacy school. It is common for each of these "roles" to have slightly different data access rights or user privileges within an electronic health record [e.g., the ability to write and sign orders for medications is usually allowed only by clinicians with a medical degree (MD, DO) or advanced nursing certification].

Advice nurse
Allergist
Allied health personnel
Anesthesiologist
Attending physician
Biomedical informatician
Biomedical informaticist
Board certified
Cardiologist
Caregiver
Case manager
Certified nurse aide (CNA)
Certified registered nurse anesthetist (CRNA)
Chief executive officer (CEO)
Chief health informatics (information) officer (CHIO)
Chief information (informatics) officer (CIO)
Chief information security officer (CISO)
Chief medical informatics (information) officer (CMIO)
Chief nursing informatics (information) officer (CNIO)
Chief operating officer (COO)
Chief quality and informatics (information) officer
Clinical champion
Clinical informatician
Clinical informaticist
Clinical nurse specialist (CNS)
Clinical research informatician
Early adopter
EHR super user
End user
Expert witness
Fellow
Gastrointestinal (GI) specialist

Gatekeeper
General medical practitioner (GP)
General surgeon
Genital-urinary (GU) specialist
Geriatrician
Healthcare paraprofessional
Health data broker
Health data custodian
Health informatician
Health informaticist
Health personnel
Healthcare proxy
Help at the elbow
Home health aide
Hospitalist
House staff
Immunologist
Informatician/informaticist
Intern
Internist
Interprofessional teams
Interventional radiologist
Intravenous (IV) nurse
Licensed clinical social worker (LCSW)
Licensed practical nurse (LPN)
Licensed vocational nurse (LVN)
Medical assistant (MA)
Medical student
Medical technologist (MT(ASCP))
Medical technologist in molecular pathology (MP(ASCP))
Multidisciplinary teams
Neurologist
Neurosurgeon
Nurse
Nurse anesthetist
Nurse practitioner (NP)
Nursing student
Obstetrician/Gynecologist (OB/GYN)
Occupational therapist (OT)
Orderly
Orthopedist
Parents or relatives
Pharmacist
Pharmacy technician
Physical therapist (PT)

Physician assistant (PA)
Plastic surgeon
Podiatrist
Postgraduate year (PGY) 1–8
Primary care provider (PCP)
Private duty nursing
Provider
Pulmonologist
Registered dietician (RD)
Registered nurse (RN)
Registered pharmacist (RPh)
Research informatician
Resident
Respiratory therapist (RT)
Respite care
Service class provider
Service class user
Skilled care
Stakeholder
Support groups
Surgeon
Surrogate
Trauma surgeon
User training

Clinical Specialty

A clinical specialty is a name for a particular branch of medicine. After completing their medical school training, physicians or surgeons usually further their medical education in a specific specialty of medicine by completing a multiple year residency training program and sometimes an additional multiple year fellowship training program to become a medical specialist. In most cases there are additional tests or "board examinations" that these clinicians must pass before they are able to practice as a board-certified specialist in their chosen subfield of medicine or surgery.

Adolescent medicine
Allergy and immunology
Anesthesiology
Cardiology
Clinical and laboratory medicine
Colon and rectal surgery
Critical care medicine
Cytopathology
Dermatology
Diagnostic radiology
Digital radiology
Emergency medicine
Endocrinology
Family medicine
Family practice
Forensic pathology
Forensic psychiatry
Geriatrics
Gerontology
Gynecology (GYN)
Hematology
Hyperbaric medicine
Infectious diseases (ID)
Internal medicine (IM)
Medical genetics
Microbiology
Nephrology
Neurology
Nuclear medicine
Obstetrics (OB)
Oncology
Ophthalmology
Orthopedic surgery
Orthopedics

Otolaryngology
Pain medicine
Pathology
Pediatrics
Plastic surgery
Podiatry
Preventive medicine
Psychiatry
Pulmonary medicine
Radiation oncology
Radiology
Rehabilitation services
Rheumatology
Speech therapy
Sports medicine
Surgery
Thoracic surgery
Transfusion medicine
Urology
Vascular surgery

Clinical Syndrome

A clinical syndrome describes a patient state that consists of a constellation of several medical signs, symptoms, and/or other physical or emotional characteristics that often occur together. Some syndromes, such as Down syndrome, have only one cause; others, such as Parkinsonian syndrome, have multiple possible causes. In other cases, the cause of the syndrome is unknown.

Acquired immunodeficiency syndrome (AIDS)
Acute coronary syndrome (ACS)
Acute respiratory distress syndrome (ARDS)
Andersen syndrome
Down syndrome
Menopause
Premenstrual syndrome (PMS)
Severe acute respiratory syndrome (SARS)
Shock
Spell
Stockholm syndrome
Sudden infant death syndrome (SIDS)
Systemic inflammatory response syndrome (SIRS)
Tetralogy of Fallot
Vertigo
Wolf–Hirschhorn syndrome

Communication

The act or process of using mutually agreed upon words, sounds, pictures, gestures, or behaviors to convey an intended meaning (e.g., thoughts, feelings, findings, or ideas) from one group to another. There are numerous options or channels (e.g., visual, haptic, auditory, olfactory, electromagnetic, kinesics, or biochemical) in which this communication can occur. Human communication is unique and often open for numerous interpretations due to its extensive use of abstract language constructs involving words, signs, symbols, or sounds.

Acknowledgment
Asynchronous
Body of message
Channel
Channel capacity
Header of message
Isochronous
Listserve
Mailing list
Public area branch exchange
Public switching telephone network
Receiver
Sender
Signal-to-noise ratio
Situation, Background, Assessment and Recommendation (SBAR) technique
Social contagion
Social network
Spam
Spamming
Synchronous communication
Transaction set
Voicemail

Computational Algorithm

A computational algorithm (pronounced AL-go-rith-um) is an unambiguous set of steps, a procedure, or a formula a computer can use to perform a specific task or solve a problem. Algorithms can be expressed in any language, including natural languages such as English, French, or Spanish to advanced programming languages such as Perl, C++, or Java. A computer uses algorithms to solve specific problems. There can be many different algorithms to solve the same type of problem. The most "elegant" algorithms often have the fewest steps, execute the fastest, and use the least amount of computer memory.

AdaBoost
Algorithm accuracy evaluation
Algorithm performance, space
Algorithm performance, time (big O)
Apriori algorithm
Artificial neural networks (ANN)
Association rule learning algorithm
Averaged one-dependence estimators (AODE)
Backpropagation
Basic Local Alignment and Search Technique (BLAST)
Bayesian algorithm
Bayesian belief network (BBN)
Binary search
Boosting
Bootstrapped aggregation (bagging)
Breadth-first search
Bubble sort
C4.5 and C5.0 (different versions of a powerful approach)
Chi-squared automatic interaction detection (CHAID)
Classification and regression tree (CART)
Collaborative filtering
Computational complexity
Conditional decision trees
Convolutional neural network (CNN)
Crowdsourcing
Cryptographic hashing functions
Data compression algorithm
Decision stump
Decision tree algorithm
Deep belief networks (DBN)
Deep learning algorithm
Deep Boltzmann machine (DBM)
Depth-first search

Dimensionality reduction algorithm
Eclat algorithm
Elastic Net
Ensemble algorithm
Evolutionary algorithm
Exhaustive search
Expectation maximization (EM)
Feature selection algorithm
Filtering algorithm
Finite-state machine
First-order predicate logic
Flexible discriminant analysis (FDA)
Fourier transform
Gaussian Naive Bayes
Generalized linear models
Genetic algorithms
Gradient boosted regression trees (GBRT)
Gradient boosting machines (GBM)
Hash function
Hierarchical clustering
Hopfield network
Huffman coding
Insertion sort
Instance-based algorithm
Iterative Dichotomiser 3 (ID3)
k-Means
k-Medians
k-Nearest Neighbor (kNN)
Learning vector quantization (LVQ)
Least absolute shrinkage and selection operator (LASSO)
Least-angle regression (LARS)
Lift
Linear discriminant analysis (LDA)
Linear regression
Locally estimated scatterplot smoothing (LOESS)
Locally weighted learning (LWL)
Lossless compression
Lossy compression
M5
Markov cycle
Markov model
Markov process
Merge sort
Mixture discriminant analysis (MDA)
Multidimensional scaling (MDS)

Multinomial Naive Bayes
Multivariate adaptive regression splines (MARS)
Naive Bayes
Neural network
NP hard
Numerical methods
Ordinary least squares regression (OLSR)
Partial least squares regression (PLSR)
Perceptron
Performance measures
Principal component analysis (PCA)
Principal component regression (PCR)
Probabilistic matching algorithm
Projection pursuit
Proxy calculations
Quadratic discriminant analysis (QDA)
Quick sort
Radial basis function network (RBFN)
Random forest
Recursive algorithms
Refinement
Regression algorithm
Regularization algorithm
Reinforcement learning
Ridge regression
Sammon mapping
Seasonal and Trend decomposition using Loess (STL decomposition)
Secure Hash Algorithm 1 (SHA-1)
Secure Hash Algorithm 2 (SHA-2)
Self-organizing map (SOM)
Semisupervised learning
Stacked auto-encoders
Stacked generalization (blending)
Stepwise regression
Supervised learning
Support vector machine (SVM)
t-Distributed Stochastic Neighbor Embedding (t-SNE)
Training data set
Transpose
Unsupervised learning
Verhoeff algorithm
Viterbi algorithm

Computer Application

An application is a computer program, or group of interacting programs, that perform a set of coordinated tasks to help the user. Applications run inside of the computer's operating system software. Applications designed for desktop or laptop computers are referred to as desktop applications. Applications built specifically for mobile computing platforms are often called apps.

AI-Rheum
Antibiotic Assistant Program
Armed Forces Health Longitudinal Technology Application (AHLTA)
Attending
Automated Medical Record (AMR)
Backwards compatibility
Bar Code Medication Administration (BCMA)
Behavioral Risk Factor Surveillance System (BRFSS)
Best-of-breed
Billing System
Biomed Central
Blue Button
Brigham & Women's Integrated Computing System (BICS)
Browser
Citation manager
Clinical data registry
Clinical data repository (CDR)
Clinical documentation
Clinical Image Access Service (CIAS)
Clinical information system (CIS)
Clinical Observation Access Service (COAS)
Clinical Trials Management System (CTMS)
ClinicalTrials.gov
Coaching expert system
Common Object Request Broker Architecture (CORBA)
Composite HealthCare System II (CHCS–DoD)
Computational propaganda
Computer-Assisted Patient Interviewing (CAPI)
Computer-Based Training (CBT)
Computer program
Computer-Stored Ambulatory Record System (COSTAR)
Computer-based Patient Record System (CPRS)
Computerized Patient Record (CPR)
Computerized Physician/provider Order Entry (CPOE)
Continuous speech recognition
Control system

Custom-designed system
Data acquisition
Data compression
Data processing
Data recording
Data transcription
Data transformation
Database management
Database management system (DBMS)
Debugger
Decision-support system
Departmental system
Disease registry
DxPlain
e-Consent
Electronic Health Record (EHR)
Electronic mail (e-mail)
Electronic medical record (EMR)
Electronic Medication Administration Record (eMAR)
Electronic nursing record
Electronic Patient Record (EPR)
Electronic Transmission of Perscription (ETP)
Enterprise Information System (EIS)
Enterprise Master Patient Index (EMPI)
Expert system
Fraud detection
Front-end application
General regular expression parser (GREP)
Geographic Information Systems (GIS)
Gmail
Graph database
Groupware
Guidance
Hadoop
Health Evaluation through Logical Programming (HELP)
Health Information Exchange (HIE)
Helper app
Homepage
Hospital Information System (HIS)
Iliad (Diagnostic Decision Support System)
Image recognition
Immunization registry
Inference engine
Information processing
Integrating Information from Bench to Bedside (I2b2)

Interactive Voice Response (IVR)
Interface engine
Internet Browser
Internist-1
Inventory management
Kaggle
Knowledge base system
LaTex
Longitudinal medical record
Management Information System (MIS)
Map reduce
Master Patient Index (MPI)
Master Provider File (MPF)
Medlars online (Medline)
MedlinePlus
MedWeaver
Metathesaurus
Mosaic browser
Mycin
Newsgroup
Niche vendor
Nursing information system
Object-oriented database
Oncocin
OPAL
Optical character recognition (OCR)
Order entry
Order entry system
Pathfinder
Patient care system
Patient portal
Patient tracking application
Patient-centered Access to Secure Systems Online (PCASSO)
Pediatric Early Warning Score (PEWS) system
Personal Health Record (PHR)
Personally controlled health management system
Personally controlled health record
Pharmacy information system
Picture Archiving and Communication (PACS)
Plugin
Point of care system
Practice management system
Problem-oriented Medical Record System (PROMIS)
PRODIGY
PROforma

Protégé
Prototype system
Provider profiling system
PubMed
Question answering programs
Quick Medical Record (QMR)
Recommendation engine
Red Cap
Relational Data Base Management System (RDBMS)
Relational database
Report Program Generator (RPG)
Results review
Rule-based expert system
Search engine
Search technology
Siri
Skype
Social media
Specialized registry
Speech recognition
Speech understanding
Spreadsheet
Statistical package
System programs
Technicon medical information system (TMIS)
The Medical Record (TMR)
Third-party
TRICARE Online
Turnkey system
Vaccine Adverse Event Reporting System (VAERS)
Value-added reseller (VAR)
Vista
Voice recognition
Web BLOB Service (WBS)
Web browser
Web catalog
Web crawler
Web-Based Training (WBT)
Wizorder
Word processor

Computer Architecture

A computer's architecture provides a framework for the rules that describe the capabilities, functionality, organization, and sometimes the methods of implementing various types of applications or computer systems.

Application program
Applications design
Architecture (computer, information)
Archival storage
Batchmode
Business logic layer
Central computing system
Centralized database
Client/server architecture
Complex Instruction Set Computing (CISC)
Computer architecture
Data layer
Distributed Component Object Model (DCOM)
Distributed data architecture
Dynamic Data Exchange (DDE)
Emergency Data Sets Framework (EDSF)
Federal Health Architecture (FHA)
Federated model
Health informatics Service Architecture (HISA)
Health information access layer (HIAL)
High-level process
Integrated versus interfaced
Java Database Connectivity (JDBC)
Lexicon query service (LQS)
Low-level process
Massive Parallel Processing (MPP)
Medicaid Information Technology Architecture
Middleware
Modular computer system
Multiprocessing
Multiuser system
National Information Infrastructure (NII)
Network-based hypermedia
Online Transaction Processing (OLTP)
Open system
Parallel processing
Patient Identification Services (PIDS)
Presentation layer
Reduced Instruction Set Computing (RISC)

Reference architecture
Reference Model for Open Distributed Processing
Regulated clinical research information model
Remote Job Entry (RJE)
Remote Procedure Calls (RPC)
Representational State Transfer (REST)
Scalability
Sequential Access Method (SAM)
Service-oriented Architecture (SOA)
Simple Object Access Protocol (SOAP)
Single user system
System
Systems aggregation
Systems integration
Terminal server
Terminate and Stay Resident (TSR)
Timesharing mode
Turing machine
User interface layer
von Neumann machine
Very Large Scale Integration (VLSI)
View schemas
Virtual Storage Access Method (VSAM)
Visual Basic Architecture (VBA)
Web Access to DICOM-persistent Objects (WADO)
Workflow engine
World Wide Web (WWW)

Computer Hardware

Computer hardware, often referred to as hardware when discussing computer-related topics, are the physical elements used to create a functional computer system, such as the microprocessor, memory, network, monitor, keyboard, data storage, all of which are tangible physical objects. By contrast, software is the set of instructions that can be stored and run by hardware to complete a task.

Analog computer
Application service provider (ASP)
Backup electrical generator
Cable
Cathode ray tube (CRT)
Central monitor
Central processing unit (CPU)
Client
Cloud computing
Compact disk (CD)
Compact disk read-only memory (CD-ROM)
Computer on Wheels (COW)
Computer system
Data bus
Deactivate
Digital computer
Digital subscribe line (DSL)
Digital video disk (DVD)
Direct-access storage device (DASD)
Display monitor
Distributed computing system
Dynamic random-access memory (DRAM)
Electronic Numerical Integrator and Computer (ENIAC)
Environment (computing)
Exam room computers
File server
Flash card
Flash memory
Floppy disk
Handheld device
Hard disk
High performance computing (HPC)
Hot site backup
Ink-jet printer
Integrated circuit (IC)
Laptop computer

Laser printer
Life cycle
Liquid crystal display (LCD)
Macintosh
Magnetic disk
Magnetic tape
Mainframe computer
Medical information bus (MIB)
Memory
Memory stick
Microchip
Modulator-demodulator (MODEM)
Netbook computer
Network protocol
Off-line device
Online device
Optical disc
Original equipment manufacturer (OEM)
Patient monitor
Personal computer (PC)
Personal digital assistant (PDA)
Physicians' workstation
Print server
Printer
Product
Random-access memory (RAM)
Raster scan display
Read-only memory (ROM)
Read-only backup
Reboot (computer)
Red electrical outlet/plug
Redundant array of independent (inexpensive) disks (RAID)
Scanning devices
Server
Smartphone
Star topology
Storage
Switch
System integration
System interface
Telemedicine technologies
Terminal
Terminal interface processor
Test system
Thick client

Thin client
Transmitter
Twisted-pair wires
Uninterruptible power supply (UPS)
Universal workstation
Variable memory
Video display terminal (VDT)
Virtual memory
Volatile memory
Warm site backup
Workstation
Workstation-on-wheels (WOW)
Write it once system
Write once read many (WORM)

Computer Networking

The use of computers and associated hardware to create a telecommunications network that can be used to facilitate the exchange of data, information, or services among individuals, groups, or institutions. Computer networks often differ in their transmission medium (e.g., copper wires, fiber optics, radio frequencies, or microwaves) used to carry their signals, communications protocols to organize network traffic, the network's size, topology, and organizational intent. In most cases, application-specific communications protocols are layered (i.e., carried as payload) over other more general communications protocols.

127.0.0.1 (localhost)
Advanced Research Project Agency Network (ARPANET)
Asymmetric digital subscriber line (ADSL)
Asynchronous Transfer Mode (ATM)
Backbone network
Bandwidth
Baud rate
Bit rate
Bits per second
Bluetooth
Broadband network
Broadband signal
Broadband transmission
Circuit switched network
Citrix
Coaxial cable
Communication protocol
Computer network
Cyberspace
Daisy chain networking
Dial tone multifrequency (DTMF)
Domain
Dynamic DNS (domain name service)
Ethernet
Fiber-optic cable
Frame relay
Gateway
Gigabit per second (Gbps)
Global system for mobile communications (GSM)
Hyperlink
Information super highway
Integrated Delivery System/Network (IDS) (IDN)
Internet
Internet relay chat (IRC)
Intranet

IP address
Latency
Local area network (LAN)
Megabit
Megabits per second (Mbps)
Microwave
Name authority
Name server
National Health Information Infrastructure (NHII)
National Health Information Network (NHIN)
Network
Network access provider
Network bridge
Network latency
Network node
Network operating system
Network router
Network services
Network stack
Network topology
Next-generation Internet
Node
Open Systems Interconnection (OSI) seven layer model
Packet
Packet-switched network
Peer-to-peer networking
Private branch exchange (PBX)
Proxy
Remote access
Remote presence health care
Router
Secure hypertext transfer protocol (SHTTP)
Store and forward
Subnet
System administration
Telecommunication
Telepresence
Token ring ethernet
Transmit (XMT)
Transmittal (XMTL)
Uniform resource identifier (URI)
Uniform resource locator (URL)
Uniform resource name
Wide area network (WAN)
Wi-Fi (Wireless Infrastructure)
WiFi Spectrum

Computer Programming

Computer programming (or just programming) is a process that leads from the initial formulation of a problem that the computer can help solve through the intricate process required to create an executable computer program. The programming process involves activities such as analysis of the problem or entire business, developing understanding of the tasks to be accomplished and the existing workflow, generating algorithms required to manipulate the data elements required to solve the problem, verification of requirements of the chosen algorithms including their appropriateness, correctness, computational resource consumption, and implementation of these algorithmic concepts (commonly referred to as coding) in the chosen programming language. The purpose of programming is to find a sequence of instructions that will automate performing a sequence of specific tasks or solving a given problem. The process of programming thus often requires expertise in many different subjects, including knowledge of the application domain, specialized algorithms and formal logic.

Active storage
Addition
Agile
Agile coach
Ajax
Alphabetic ordering
Analog-to-digital conversion (ADC)
Android
Applets
Application Programming Interface (API)
Apps
Assembler
Assembly code
Binary sort
Bit (short for binary digit)
Boot
Buffer
Buffer overflow
Burn down
Business logic
Closed loop control
Code
Code review
Coercion

Command
Compiler
Compiler optimization
Computed check
Computer bug
Computer-readable content
Concept modeling
Consistency check
Constraint-based programming
Construction
Daily standup
Data accessibility
Data architecture
Data capture
Data control structure
Data element
Data flow
Data flow diagram
Data independence
Data mediator
Data model
Data quarantining
Data set
Data storage
Data stream
Database
Database recovery
Debug
Delta check
Demonstration (demo)
Design by constraint
Division
Document Type Definition (DTD)
Dynamic programming
Entity, attribute, value (EAV)
Entity–Relationship diagram (ER or ERD)
Error trap
Exception handling
Extended Architecture Operation System (XA)
Floating point exception
Generalization
Global variable
Hash table
Hashing

Hierarchical database
Information Object
Information Object Class
Information Object Instance
Input
Integrated Development Environment (IDE)
Interpreter
Iteration
Iterative
Java 2 Platform, Enterprise Edition (J2EE)
Job
jQuery
Kernel
Late binding
Linux
Local variable
Machine code
Macro
Markup
Marshalling
Mathematical operations
Model View Controller (MVC)
Multiplicity
Multiprogramming
Node.js
Object
Object modeling
Object oriented programming (OOP)
Object-based approach
Object-oriented analysis
Object-oriented programming
Open loop control
Output
Page
Pattern check
Pointer-to-data
Product backlog
Product owner
Regular expression
Remote Method Invocation (RMI)
Requirements development process
Resource Definition Format (RDF)
ReST Protocol
Retrospective

Rounding error
Scrum
Semantic web
Serialization
Service
Shell script
Simultaneous access
Simultaneous controls
Software
Software assurance
Software design patterns
Software engineering
Source
Source code compartment
Specialization
Sprint
Sprint backlog
Sprint planning
Sprint review
Structured programming
Stub
Style sheets
Subtraction
Synchronization
System specification
Systems development
Team velocity
Technical specifications
Termination
Text editor
Truncate
Type checking
Unified Modeling Language (UML)
Unix
User stories
Virtual addressing
Virtualization
Waterfall model
Website design
Windows
Word
Word size
Working memory
Wrapper

Computer Security

The protection of computers and the computing infrastructure (i.e., hardware, software, data, information, or knowledge) from theft, damage, disruption, or misdirection of services it provides. It includes protecting and controlling physical access to the hardware, as well as protecting the software or data from harm that may come via inappropriate network access, data corruption, or code injection due to malicious activities by internal or external agents, whether intentional, accidental, or due to someone being tricked into deviating from routine security procedures.

Acceptable use policy
Access
Access control mechanism
Active content
Administrative safe guards
Advanced persistent threat
Adversary
Anonymization of data
Asset
Availability
Backup
Behavior monitoring
Bioterrorism
Bootkit
Box
Business continuity
Check digit
Checksum
Cipher
Ciphertext
Code escrow
Completely Automated Public Turing test to tell Computers and Humans Apart (Captcha)
Computer forensics
Computer security incident
Consequence
Continuity of operations plan
Cryptanalysis
Cryptographic algorithm
Cryptographic encoding
Cryptography
Cryptology
Cyber ecosystem
Cyber exercise

Cyber incident
Cyber incident response plan
Cyclic redundancy checks
Data availability
Data breach
Data encryption
Data encryption standard
Data integrity
Data leakage
Data loss
Data lost prevention
Data privacy
Data redundancy/mirroring
Data reidentification
Data spill
Data theft decipher
Dated administration
Decode
Decrypt
Decryption
Digital forensics
Disaster recovery
Disaster recovery plan (DRP)
Downtime
Encipher
Encode
Encrypt
Encrypted
Encryption
Enterprise risk management
Event
Exfiltration
Exploit
Exploitation analysis
Exposure
Failure
Fletcher's checksum
Identification (ID)
Incident management
Incident response
Incident response plan
Information assurance
Information assurance compliance
Information communication technology (ICT) supply chain
threat

Information security policy
Information sharing
Information system resilience
Information system security operations
Information Technology (IT) asset
Insider threat
Integrated risk management
Integrity
Intelligence, surveillance, and reconnaissance (ISR)
Intense
Internet Security Assessment (ISA)
Investigate
Investigation
Keep pair
Keystroke logger
Lots of Copies Keep Stuff Safe (LOCKSS)
Machine learning and evolution
Master boot record
Minimum necessary data set
Mitigation
Moving target defense
Multilevel security
Operational exercise
Outsider threat
Passive attack
Physician identification number (PIN)
Plaintext
Precursor
Preparedness
Privacy
Protected health information (PHI)
Provider identification number (PIN)
Recovery
Redundancy
Reidentification
Resilience
Response
Response plan
Risk analysis
Risk assessment
Risk mitigation
Risk-based data management
Rootkit
Secret key
Secret key cryptography

Secure Hash Standard (SHA)
Secure shell (ssh)
Security architecture and policy
Security audit
Security automation
Security flaw
Security incident
Security policy
Security program management
Security provision
Security risk assessment
Situational awareness
Situs State
Spear phishing
Spillage
Spoofing
Symmetric cryptography
Symmetric encryption algorithm
Symmetric key
Tabletop exercise
Tailored trustworthy space
Targets threat
Terminal ID (TERMID)
Threat actor
Threat agent
Threat analysis
Threat and vulnerability assessment (TVA)
Threat assessment
Ticket
Traffic light protocol
Unauthorized access
User Datagram Protocol (UDP)
User Identification (USERID)
Vulnerability
Vulnerability assessment and management
Web widget
Whale phishing
White list
White-hat hacker
Wipe the disk

Computer-Based Education

A type of curricula in which students interact with a computer as a key element of the learning process. Students may read materials, watch or listen to recordings, complete exercises, interact with models or simulations, and discuss examples, via computer rather than receiving the information from printed materials or an instructor's oral presentation. Computer-based education is most often accomplished asynchronously, in that the instructor and students are most often not interacting or communicating with each other at the same time.

Avatar
Computer-based simulation
Conceptual fidelity
Confederate
Continuing medical education (CME)
Discrete event simulation
European Computer Driving Licence (ECDL)
International Computer Driving Licence (ICDL)
Maintenance of certification (MOC)
Manikin (mannequin)
Massive Open Online Course (MOOC)
Professional development
Simulated patient
Standardized patient
Transformation-based learning
Tutoring system
Virtual patient

Corporation

A corporation is a legal entity that is separate and distinct from its owners. Corporations enjoy most of the rights and responsibilities that an individual possesses. For example, a corporation has the right to enter into contracts, loan and borrow money, sue and be sued, hire employees, own assets and pay taxes. It is often referred to as a "legal person." This means that the corporation itself, not the people who make it up or the people who own it, is held legally liable for the actions and debts the business incurs.

Allscripts
Aprima Medical Software
AthenaHealth
Cerner Corporation
Computers Programs and Systems, Inc. (CPSI)
eClinicalWorks (eCW)
ECRI
E-MDs
Epic Systems Corporation
General Electric (GE) Health care
Greenway Medical Technologies
Hospital Corporation of America (HCA)
International Business Machines (IBM)
Meditech
NextGen Healthcare Information Systems Inc.
Practice Fusion
Red Hat
Surescripts
Telco

Data Analysis

A systematic process for collection of raw data, inspecting, cleaning, transforming, and modeling that data with the goal of turning it into useful information, suggesting conclusions, and supporting decision-making.

Data variability
Data variety
Data velocity
Data veracity
Data volume
Digital signal processing (DSP)
Drill-down analysis
Frequency modulation
Garbage in, garbage out (GIGO)
Instrumental variable
Pattern recognition
Pharmacokinetic parameters
Pharmacovigilance
Protocol analysis
Waveform template
Wavelet compression

Data Structure

A manner in which data can be organized in a computer so that they can be used efficiently in an algorithm or for analysis. A data structure is a concrete implementation of a specific abstract data type. Different kinds of data structures are suited to different kinds of applications, and some are highly specialized to specific tasks. Data structures provide a means to manage large amounts of data efficiently.

Address (data)
Associative array
Binary tree
Data arrays
First in, first out (FIFO)
First in, last out (FILO)
Graph
Linked list
Sets
Stack (data)
Tree

Data Type

The term data type is used in different scientific contexts to refer to the methods of classifying data according to the possible values for that type, the operations (e.g., statistical methods) that can be done on values of that type, the meaning of the data, and (especially in computer science) the way values of a particular type of data can be stored.

9-digit ZIP Code Plan (ZIP + 4)
Absolute time
Alphabet
Alphanumeric
Analog data
Analog signal
Array
Array list
Bit array
Bitmap
Boolean (true or false)
Cartesian tree
Character
Circular buffer
Coded data
Common Clinical Data Set
Container
Continuous data
Control table
Data
Datum
Delimited character string
Digital data
Digital signal
Discriminated union
Disjoint union
Double floating point
Double-ended queue
Doubly linked list
Dynamic array
Enumerated
Floating point
Free list
Freetext
Genetic data
Geospatial data

Greenwich mean time (GMT or ZULU)
Gregorian date
Hashed array tree
Iliffe vector
Image data
Integer
Interval scale
Irrational number
Julian date
List
Lookup table
Medical data
Multidimensional data
Multimap
Multiset
Noise
Nominal scale
Nonnumeric characters
Ordinal scale
Outcomes data
Parallel array
Patient specific information
Patient-generated data
Pixel
Priority queue
Queue
Randomized binary search tree
Ratio scale
Rational number
Real number
Record
Relative time
Self-balancing binary search tree
Serial data
Set
Sorted array
Sparse array
Sparse matrix
Structured content
Tagged union
Telemetry
Tuple
Unstructured data
Value set

Variable-length array
Variant record
Vector
Weight-balanced tree
XOR linked list
Year (YR)
Year (YYYY)
Year 2000 (Y2K)
Year of birth (YOB)
Year-to-date (YTD)

Data Visualization

Data visualization involves the graphical display of data to facilitate its analysis and communication by making it more accessible, understandable, and usable. Data visualization can be both an art and a science. To convey ideas effectively, the data's aesthetic form and functionality must be accurately portrayed.

Abscissa
Amplitude
Asymptote
Augmented reality
Bar chart
Bar graph
Baseline
Boundary conditions
Box and whiskers plot
Bubble chart
Data display
Flowcharts
Forest plot
Graph
Graphical analysis
Line
Logarithmic scale
Minimum
Monotonic
Ordinate
Orthogonal
Overshoot
Peak
Plateau
Rate of change
Sinusoidal waveform
Slope
Spike
Star plot
Time course
Trend
Trough
Undershoot
Waveform
Whiskers plot

Data Warehousing

The process of creating a central repository of data often uploaded from multiple disparate operational data sources that can be used for data analysis and reporting. The data are often transformed from the operational or transactional systems and integrated (or combined) to facilitate different types (e.g., longitudinal) of queries.

Administrative versus clinical data
Aggregate content
Appliances
Appropriate field size/type for data
Atomicity
Atomicity, consistency, isolation, durability (ACID)
Attribute
Batch mode processing
Business intelligence
Byte
Charge master
Chron jobs
Clean the data
Conform to mention
Consistent, standardized internal data naming
Content structuring
Data aggregation
Data attribute
Data consistency
Data consolidation (reduction)
Data cubes
Data Description Language (DDL)
Data dictionary
Data Dictionary Definition Language (DDDL)
Data dimensions
Data Element Catalog
Data integration
Data mining
Data retention policy
Data synthesis
Data Views
Data warehouse
Database backup
Datamart
Deletes
Denormalization
Deprecate

Dimensional model
Dimensional table
Draw up
Drill across
Drill down
Drill through
Durability
Electronic Data Capture (EDC)
Electronic Data Interchange (EDI)
Enterprise Data Warehouse (EDW)
Entity
Entity Frameworks
Entity–Relationship model
Exabyte
Extract, Transform, Load (ETL)
Fact table
Field
Flag fields (binary)
Flat files
Foreign key
Function graphs
Functional programming
Gantt chart
Geospatial maps
Gigabyte (Gb)
Global Unique Identifiers as primary keys (GUIDs as PKs)
Graphical Query Language (GQL)
Heat map
Hierarchy
Histogram
Inner join
Inserts
Isolation
Joins
Key field
Level of data normalization (first-, second-, third-level normalization)
Line graph
Longitudinal query
Maintenance of raw data after cleansing
Many-to-many relationship
Master Provider Index (MPI)
Megabyte (MB)
Metadata
Model organism database

Multidimensional OLAP
Nightly download
NoSQL (not only SQL)
Null
Off-site storage
One field–one concept
One version of the truth across the enterprise
Online analytical processing (OLAP)
Open Data Base Connectivity (ODBC)
Outer join
Performance benchmarks
Petabyte (Pb)
Pie chart
Postgenomic data base
Primary key
Query tuning/optimization
Query-by-example
Radar plot
Read-only access/privileges
Real-time data upload
Referential integrity
Report date
Research Data Repository (RDR)
Rolling benchmark calculation
Scatter plot
Schema
Secondary use of data
Slice and dice
Snowflake schema
Sparklines
Table relationship mapping
Tables
Terabyte (Tb)
Time series plot
Timelines
Transactional system
Tree map
Trigger
Twinkling database
Updates
Value
Virtual data warehouse (VDW)
Warehousing
X–Y plots
Yottabyte

Disease

Refers to any condition that impairs the normal functioning of the human body. Diseases are often associated with some type of dysfunction within the body's normal homeostatic process. Commonly, the term disease is used to refer specifically to infectious diseases, that result from the presence of pathogenic microbial agents, including viruses, bacteria, fungi, protozoa, multicellular organisms, and aberrant proteins known as prions. There are also noninfectious diseases, including most forms of cancer, heart disease, and genetic disease. Four main types of diseases are typically considered: pathogenic diseases, deficiency diseases, hereditary diseases, and physiological diseases.

Acute bronchitis
Acute disease
Acute illness
Anthrax
Appendicitis
Bacterial sepsis
Black lung (pneumoconiosis)
Chronic lower respiratory disease
Cytomegalovirus (CMV)
Diphtheria
Epidemic
Gastroesophageal reflux disease (GERD)
Graft versus host disease (GVHD)
Group B strep
Haemophilus influenza
Heart disease
Hepatitis B (Hep B)
Hepatitis C virus (HCV)
Human immunodeficiency virus (HIV)
Human papillomavirus (HPV)
Influenza (flu)
Ischemic vascular disease
Japanese encephalitis (JE)
Liver disease
Measles
Meningitis
Meningococcal infection
Methicillin-resistant Staphylococcus aureus (MRSA)
Mumps
Necrotizing enterocolitis (NEC)
Nephritis
Pathogen

Pathological
Peptic ulcer
Pertussis
Pneumococcal pneumonia
Pneumonia
Pneumonitis
Polio
Rabies
Reportable diseases
Respiratory distress
Respiratory syncytial virus (RSV)
Rotavirus
Rubella
Septicemia
Sexually transmitted disease (STD)
Shingles
Smallpox
Symptoms (Sx)
Systemic lupus erythematosus (SLE)
Tetanus
Tuberculosis (TB)
Upper respiratory infection (URI)
Urinary tract infection (UTI)
Varicella
Venereal disease (VD)
Viral hepatitis
Virus
Yellow fever
Zika virus

Electronic Health Record Function

An electronic health record is a compilation of software routines that provide all the features and functions (e.g., data capture, order creation, information sharing, recording clinician findings, thoughts, and actions, and storage of an accurate and complete copy of a patient's health record) required to help clinicians (e.g., physicians, nurses, respiratory therapists, nutritionists, etc.) practice medicine.

Accessibility-centered design
Accounting of disclosures
Active order
Active problem
Activities of daily living (ADLs)
Admission Discharge Transfer (ADT)
Advance Health Care Directive
After Visit Summary (AVS)
Amendments
Application access—all data request
Archive
Assessment
Audit report(s)
Auditable events and tamper-resistance
Automated measure calculation
Automated numerator recording
Automatic access time-out
Baseline rate, population
Bed Master File (BED)
Best Practice Alert (BPA)
Break The Glass (BTG)
Cancelled order
Chief Complaint (CC)
Clinical information reconciliation and incorporation
Clinical practice guidelines
Clinical quality measures (CQMs)—record and export
Common Clinical Data Set summary record—create and receive
Contact information
Cosign
Coverage list
Data export
Data segmentation for privacy
Date of birth (DOB)
Date of death (DOD)
Date/Time stamp

Demographics
Dependent
Dictation
Differential diagnosis
Direct Project
Discharge summary
Discontinued (DC) order
Do Not Resuscitate (DNR) order
Dose
Drug-formulary and preferred drug list checks
Edge Protocol
Electronic Prescribing (eRx)
Electronic Reportable Lab
Electronic signature (eSignature)
Emergency access
Encounter-based record
End-user device encryption
Episode-based record
Estimated Date of Confinement (EDC)
Family health history
Filled prescription
Frequency of administration
Hand-off
Health Risk Assessment (HRA)
History (hx)
History of Present Illness (HPI)
Implantable device list
Laboratory information system (LIS)
Medical history for all children
Medication allergy list
Medication history
Medication list
Medication route
No Known Allergies (nka)
Objective
Order catalog
Order Entry (OE)
Ordering provider
Parts of order
Past Medical History
Patient health information capture
Patient-specific education resources
Pending order
Perscription
Physician Orders for Life-Sustaining Treatment (POLST)

Plan
Problem list
Problem-oriented medical record (POMR)
Profile
Psychiatric history
Quality system management
Radiology Information System (RIS)
Reason for referral
Referral
Resolved problem
Results reporting
Review of Systems (ROS)
Safety-enhanced design
Secure messaging
Sign & Hold (S&H)
Signature
Sign-out
Smoking history
Smoking status
Social history
Social, psychological, and behavioral data
Stop date
Subjective
Subjective, Objective, Assessment, Plan (SOAP) note
Surgical history
Test name
Transmission to cancer registries
Transmission to immunization registries
Transmission to public health agencies—electronic case reporting
Trusted connection
Vendor system

Evaluation

A systematic set of methods for making a judgment or assessment of a subject's merit, amount, worth, number, value, or significance. These judgments are made using criteria governed by a set of standards. Formative evaluations can be used to assess an intervention, initiative, person, project, program, or even an entire organization, and help with decision-making designed to make something better. Summative evaluations can be used to ascertain the degree of achievement or value in regard to the aim and objectives and results of any such action or intervention that has been completed. The primary purpose of evaluation, in addition to gaining insight into prior or existing initiatives, is to enable reflection and assist in the identification of future decisions or change.

Cost–benefit analysis
Cost effectiveness threshold
Formative assessment
Formative evaluation
Goal-free approach
Internal validation
Marginal cost-effectiveness ratio
Objectivist
Process measure
Quasilegal approach
Responsive-illuminative approach
Staged evaluation
Subjectivist
Summative assessment
User profiling
Validation
Web analytics

Field of Study

A branch of knowledge that is taught and researched as part of higher education. A student or scholar's field of study, or academic discipline, is commonly defined and recognized by university faculties, learned societies, and the academic journals that publish research in that particular scientific area. In general, the specific knowledge that is included in any specific academic discipline is open to debate and commonly, multiple fields of study cover the same knowledge.

Anthropology
Anthropometry
Artificial intelligence (AI)
Basic research
Basic science
Behavioral economics
Behaviorism
Bibliometrics
Big data
Bioethics
Bioinformatics
Biomedical computing
Biomedical engineering
Biomedical informatics
Biostatistics
Calculus
Causal modeling
Clinical informatics
Clinical research
Clinical research informatics
Cognitive science
Cognitive work analysis
Comparative effectiveness research (CER)
Complementary and alternative medicine (CAM)
Complexity science
Computational linguistics
Computational intelligence
Computer programming
Computer science (CS)
Computer supported cooperative work (CSCW)
Computer vision (CV)
Computer-aided instruction (CAI)
Computer-based education
Consumer health informatics

Cybernetics
Data science
Data visualization
Database design and administration
Decision analysis
Decision science
Decision support
Dental informatics
Descriptive statistics
Distributed cognition
eHealth
Epidemiology
Ergonomics
Ethnography
Evidence-based medicine (EBM)
Experimental science
Genomics
Health informatics
Health information management (HIM)
Health information technology (HIT)
Health policy
Health services research
Health technology assessment (HTA)
Human–computer interaction (HCI)
Human factors
Imaging informatics
Implementation science
Industrial engineering (IE)
Inferential statistics
Informatics
Information and communications technology (ICT)
Information science
Interventional radiology
Lexicography
Linear systems
Machine learning
Medical anthropology
Medical computer science
Medical computing
Medical decision making
Medical informatics
Medical information science
Medical management
Medical technology
Metrology

Mobile health (mHealth)
Morphology
Morphometrics
Nanotechnology
Neural computing
Neural informatics
Nonlinear systems
Nosology
Nuclear medicine imaging
Numerical analysis
Nursing informatics
Operations research
Outcomes research
Pathophysiology
Persuasive technology
Population health
Precision medicine
Predicate calculus
Predictive modeling
Program evaluation
Public health
Public health informatics
Recommender systems
Scientific writing
Sociotechnical systems
Spatiotemporal analytics
Structural informatics
Technology assessment
Teleconsultation
Teledermatology
Telehealth
Telemedicine
Telepathology
Teleradiology
Telerobotics
Topology
Translational bioinformatics
Ultrasound imaging
Vectorcardiography
Virtual reality

Genetics

A biological field concerned with genes, heredity, and variation in living organisms. It focuses mainly on the study of the subcellular properties (i.e., molecular structures, functions of genes, and gene behavior) of cells or organisms that enable the transfer, or in some cases the inability to transfer, various traits from parents to their offspring. This transfer allows for the propagation of certain anatomical and physiological characteristics from one generation to the next.

Allele
Alternative splicing product
Base pair
Biomarker
Candidate gene study
Chromosome
Complementary DNA
Deoxyribonucleic acid (DNA)
DNA sequencing
Drive
Enzyme
Epigenetics
Fitness landscape
Functional genomics
Gene
Gene prediction
Gene product
Gene therapy
Genome
Genome level characters
Genome wide association study (GWAS)
Genotype
Horizontal gene transfer
Human Genome Project (HGP)
Hybridization
Infectome
Ligand
Messenger RNA
Microbiome
Mouse model
Mutation
Next-generation sequencing
Northern blot
Oligonucleotide

Open reading frame
Orthologous
Phage
Pharmacogenetics
Pharmacogenomics
Phenotype
Polymerase chain reaction (PCR)
Polymorphism
Proband
Protein sequence database
Proteomics
Reading frame
Ribonucleic acid (RNA)
Sequence alignment
Sequence information
Single nucleotide polymorphism (SNP)
Southern blot
Structural alignment
Systematic classification of proteins
Transcription factor
Universal genetic code
Variants
Whole genome shotgun sequencing

Government Funding

Any financial support provided by a local, state, or federal government organization often used to fund various types of scientific research. The funding is often determined through a competitive process, in which potential projects are evaluated (often by peers) and only the most promising receive funding.

Affordable Care Act (ACA)

Alcohol, Drug Abuse, and Mental Health Services Block Grant

American Recovery and Reinvestment Act of 2009 (ARRA)

Cancer Biomedical Informatics Grid (caBIG)

Career Development Award (K Award)

Children's Health Insurance Program Reauthorization Act of 2009

Clinical Translational Science Awards (CTSA)

Early and Periodic Screening, Diagnosis, and Treatment Program (EPSDT)

Health Manpower Shortage Area (HMSA)

Home and Community-Based Waivers

Integrated Advanced Information Management Systems (IAIMS)

Medicaid (Title XIX)

Notice of Award (NOA)

Notice of Grant Award (NGA)

Principal Investigator (PI)

Program of All-Inclusive Care for the Elderly (PACE)

Request for Applications (RFA)

Social Security

Social Services Block Grant (SSBG) Services

State Medicaid Health Information Technology Plan

Supplemental Security Income (SSI)

Veterans' Disability Pension Program

Government Organization

A permanent or semipermanent organization that forms a part of the government's bureaucracy. Individual organizations are often responsible for the oversight and administration of specific government functions. A government organization may be established by national, state, or local legislative or executive branches of government. The autonomy, independence, and accountability of government organizations vary widely.

Advisory Committee for Immunization Practice (ACIP)
Agency
Agency for Healthcare Research and Quality (AHRQ)
Area Agency on Aging (AAA)
Australian Health Information Council (AHIC)
Australian Health Ministers' Advisory Council (AHMAC)
Center for Disease Control and Prevention (CDC)
Centers for Medicare and Medicaid Services (CMS)
Congressional Budget Office (CBO)
Consolidated Health Informatics (CHI)
Department of Defense (DoD)
Department of Health (DoH)
Department of Health and Human Services (HHS or DHHS)
Drug Enforcement Administration (DEA)
Equal Employment Opportunity Commission (EEOC)
European Union (EU)
Federal Advisory Committee Act (FACA)
Federal Bureau of Investigation (FBI)
Federal Communications Commission (FCC)
Federal Trade Commission (FTC)
Food and Drug Administration (FDA)
General Services Administration (GSA)
Government Accountability Office (GAO)
Health Care Financing Administration (HCFA)
Health Resources and Services Administration (HRSA)
Health Systems Agency (HSA)
Home Health Agency (HHA)
Indian Health Service (HIS)
Inspector General (IG)
Maternal and Child Health Block Grant (Programs for Children with Special Needs)
Military Health System (MHS)
Ministry of Health (MOH)
National Aeronautic and Space Administration (NASA)
National Cancer Institute (NCI)
National Center for Biotechnical Information (NCBI)

National Center for Health Services and Research (NCHSR)
National Center for Health Statistics (NCHS)
National Committee for Quality Health Care (NCQHC)
National Committee on Vital and Health Statistics (NCVHS)
National Computer Security Association (NCSA)
National Guideline Clearinghouse (NGC)
National Health Service (UK) (NHS)
National Heart, Lung, and Blood Institute (NHLBI)
National Human Genome Research Institute (NHGRI)
National Institute for Health and Clinical Excellence (NICE)
National Institute of Allergies and Infectious Diseases (NIAID)
National Institute of Child Health and Human Development
National Institute of Dental Research (NIDR)
National Institute of Diabetes, Digestive, and Kidney Diseases (NIDDKD)
National Institute of Mental Health (NIMH)
National Institute of Occupational Safety and Health (NIOSH)
National Institute on Aging (NIA)
National Institute on Drug Abuse (NIDA)
National Institutes of Health (NIH)
National Library of Medicine (NLM)
National Program for Information Technology (UK) (NPfIT)
National Science Foundation (NSF)
National Security Agency (NSA)
National Transportation Safety Board (NTSB)
Occupational Safety and Health Administration (OSHA)
Office of Civil Rights (OCR)
Office of Inspector General (OIG)
Office of Management and Budget (USA) (OMB)
Office of the National Coordinator for Health Information Technology (ONC)
Patient Centered Outcomes Research Institute (PCORI)
President's Information Technology Advisory Committee (PITAC)
Public Health Agency
Public Health Department
Public Health Services (PHS)
Regional Health Information Network (RHIN)
Regional Health Information Organization (RHIO)
Veterans Health Administration (VHA)
Veterans' Health Services Programs
Vital Statistics
World Health Organization (WHO)

Health Insurance

A financial arrangement in which a company or government agency provides a future guarantee of compensation for specified medical expenses resulting from injury, illness, or death in return for payment of an upfront premium. The insurance companies or government agencies determine the upfront premium by estimating the overall medical costs associated with the individuals in the group that it is insuring.

Adjusted Average Per Capita Cost (AAPCC)
Blue Cross/Blue Shield (BC/BS)
Braided Funding
Captive
Carrier
Catastrophic Health Insurance
Child Health Insurance Program (CHIP)
Civilian Health and Medical Program of the Uniformed Services (CHAMPUS)
Civilian Health and Medical Program of the Veterans Administration (CHAMPVA)
Competitive Medical Plan (CMP)
Consumer
Coverage
Coverage Basis
Coverage Decision
Covered Services
Current Annual Premium
Current Claimant
Custodial Care
Defined Benefit
Defined Contribution
Drug Claims Processing
Drug Risk-Sharing Arrangements
Dual Eligible
Exclusive Provider Arrangement (EPA)
Expenditure Target (ET)
Federal Employees Health Benefits Program (FEHBP)
Federal Poverty Level (FPL)
Flexible Savings Account (FSA)
Formulary
Future Purchase Option (FPO)
General Liability Claims/Losses
Guaranteed Renewal
Health Insurance Purchasing Cooperative (HIPC)
Health Plan

Health Risk Factors
Health Status
High-Risk Pool
Home Health Care Benefit Amount
Indemnity insurance
Inflation Protection Duration: Life of Policy/Certificate
Institutional Long-Term Care (ILTC)
Joint and Several Liability
Joint Underwriting Association
Lifetime Maximum Structure (LMS)
Long-Term Care Insurance (LTCI)
Maintenance Assistance Status (MAS)
Major medical insurance
Medical Necessity
Medicare Advantage
Medicare Supplement Insurance (MedSupp)
Medigap
Nursing Home Liability Insurance
Offshore Captives
Partnership Status
Policy Benefit Type
Policy Number
Preadmission Certification
Preexisting Condition
Preferred Provider Arrangement (PPA)
Premium
Private health insurance
Professional Liability Claims/Losses
Psychiatric Rehabilitation Option
Qualifying Condition
Regulated Insurance Carrier
Reimbursement
Reinsurance
Remaining Lifetime Benefits
Rent-A-Captive
Restricted-Benefit Enrollee
Risk Retention Group (RRG)
Self-Insured plans
Service Plan
Social Security Disability Insurance (SSDI)
Spend-Down
Spousal Impoverishment
Underinsured
Underwriting
Veterans' Disability Compensation Program
Workers' Compensation Program

Healthcare Finance

A branch of the field of finance that describes the processes by which patients and healthcare beneficiaries pay for medical expenses. When thinking about healthcare finance, one must consider at least three questions: How is the money raised to pay for the healthcare services? How are funds from groups of patients pooled? And, how are healthcare services paid for?

Blended Funding
Capital
Capital Expenditure Review
Capitalization
Capitation Rate
Carve Out
Catchment Area
Certificate of Need (CON)
Community Rating
Cost Containment
Cost Minimization Analysis (CMA)
Cost Neutrality
Cost of Illness Analysis (COI)
Cost of Living Adjustment/Allowance (COLA)
Cost Sharing
Cost Shifting
Cost Utility Analysis
Cost-Based Reimbursement
Cost–Benefit Analysis
Cost-Shifting
Diagnostic-related Group (DRG)
Guarantor
Insurance guarantor
Medicare (Title XVIII)
Self-pay
Uncompensated Care
Uniform billing form (UB-92)
Workman's compensation

Hospital Department

The set of organizational components commonly found in hospitals. Hospital departments provide specific diagnostic or therapeutic services to patients throughout the hospital.

Acute Care
Ancillary Services
Bone Marrow Transplant (BMT) Unit
Burn unit
Cardiac Intensive Care Unit (CICU)
Cardiovascular Intensive Care Unit (CVICU)
Care/Case Management
Critical care
Critical Care Unit (CCU)
Ear, Nose, Throat (ENT)
Emergency Department (ED)
Emergency Room (ER)
Escort Services
General surgery
Graduate Medical Education (GME)
Head, Eyes, Ears, Nose, (Mouth), and Throat (HEENT)
Health Education
Health Promotion
Hospice Care
Information Systems (IS)
Information Technology (IT)
Inpatient
Intensive Care Unit (ICU)
Intermediate Care
Labor and Delivery (L&D)
Laboratory (LAB)
Long-term care Ombudsman
Medical Intensive Care Unit (MICU)
Mental Health Services
Neonatal Intensive Care Unit (NICU)
Network Operation Center (NOC)
Nursing station
Occupational Health Services
Occupational Therapy (OT)
Ombudsman
Operating Room (OR)
Pediatric Intensive Care Unit (PICU)
Perinatal
Pharmaceutical and Therapeutic (P&T) committee
Physical Therapy (PT)

Postacute Care (PAC)
Postanesthesia Care Unit (PACU)
Pulmonary Intensive Care Unit (PICU)
Rapid Response Team (RRT)
Rehabilitation
Risk management (RM)
Skilled Nursing Care
Special Care Units
Subacute Care
Surgical Intensive Care Unit (SICU)
Tertiary care
Transportation Services
Trauma line

Human–Computer Interaction

A subfield of computer science that focuses on the design and use of computing technology that provides the interface between computing technology and the people who use it. It relies heavily on the more mature fields of cognitive science and human factors engineering. An important goal of the field is to facilitate a "dialog" between humans and computers, which is similar to that of human-to-human interactions.

Abbreviations to avoid
Affordance
Augmented reality
Autocompletion
Bitmap display
Button
Cascading style sheets
Charting by exception
Chronological order
Cognitive load
Contrast
Copy and paste
Cursor
Dashboard
Data overload
Direct manipulation
Drop-down control
Error recovery
Feedback
Flowsheet
Foreground/background color combinations to avoid
Graphic editor
Graphical models
Graphical user interface (GUI)
Growth charts
Haptic feedback
Human-readable content
Hypertext
Icon
Information overload
Interface consistency
Joystick
Keyboard
Keystroke Level Modeling (KLM)
Learnability
Light pen

Make the right thing to do, the easiest thing to do
Mandatory field
Memorability
Mental models
Menu
Metaphor
Metaphor graphics
Mouse (pointing device)
Multimedia
Passive
Perception
Postscript
Presentation
Prospective memory
Radio button
Range check
Raster image
Readability
Red, green, blue pixels
Relevant feedback
Reverse chronological order
Right information to right person at right time, so they can make right decision
Screenshot
Shared mental model
Structured data
Tab control
Tab metaphor
Tactile feedback
Tiling
Touchscreen
Trackball
Usability
Usability engineering
Use error
User experience (UX)
User interaction model
User profile
User-centered design (UCD)
Vector image
View
Visibility
Visualization
What you see is what you get (WYSIWYG)
White board

Identity Management

A broad administrative area or discipline that deals with identifying individuals in a system (such as a hospital, a healthcare delivery network, or an entire community), protecting that identify, and controlling their access to resources within that system by associating specific user rights and restrictions with the user's established identity. The goal of identity management is to ensure that the right individuals are able to access the right resources at the right times and for the right reasons.

Access and identity management
Access control
Accessibility
Accountability
Attestation
Audit trail
Authenticate
Authentication
Authenticity
Authorization
Biometric authentication
Biometric identification
Biometric identifier
Comingled records
Confidentiality
Confidentiality, integrity, availability (CIA)
Data confidentiality
Deidentification
Deidentified data
Duplicate records
Face (facial) recognition
Finger print recognition
Handwriting recognition
Inaccessibility
Internet certificate
k-Anonymity
Key
Key resource
Nonrepudiation
Palm-print recognition
Password
Password change policy
Personal identification number (PIN)
Personal identifying information

Role-based security
Role-limited access
Social Security Number (SSN)
Three factor authentication—something you know, something you have, something you are
Two-factor authentication

Imaging

The process of creating a visual representation or reproduction of an object's internal or external structure. It can be used to allow clinicians to look at the inside or outside of the human body for clues about a medical condition. A variety of machines, modalities, and techniques can create visual representations of the internal and external structures and activities of the body.

Back projection
Charge-coupled device (CCD) camera
Chest photofluorography
Chest X-ray (CXR)
Color resolution
Computed radiography
Computed tomography (CT)
Computerized axial tomography (CAT)
Contrast radiography
Contrast resolution
Contrast-enhanced ultrasound
Convolution
Deformable model
Diffuse optical imaging
Diffusion tensor imaging
Diffusion-weighted imaging
Digital image
Digital image acquisition
Digital radiography
Digital subtraction angiography
Dosimetry
Echocardiography
Edge detection
Electrical impedance tomography
Feature classification
Feature detection
Feature extraction
Filmless imaging
Fluoroscopy
Functional magnetic resonance imaging (fMRI)
Functional mapping
Global processing
Gray scale
Gynecologic ultrasonography
Histogram equalization
Image database

Image enhancement
Image generation
Image management
Image manipulation
Image processing
Imaging modality
Intravascular ultrasound
Ionizing radiation
Light
Magnetic resonance imaging (MRI)
Molecular imaging
Multimodal image fusion
Neuroimaging
Nonionizing radiation
Nuclear magnetic resonance (NMR) imaging
Nuclear magnetic resonance (NMR) spectroscopy
Obstetric ultrasonography
Phantom
Positron emission tomography (PET)
Projection
Qualitative arrangement
Radioactive isotope
Region detection technique
Resolution
Shadow graph
Single photon emission computed tomography (SPECT)
Spatial resolution
Surface based warping
Surface rendering
Template Atlas
Temporal resolution
Temporal subtraction
Three-dimensional reconstruction and visualization
Three-dimensional structure information
Ultrasound
Unsharp masking
Virtual colonoscopy
Volume rendering
Volume-based warping
Voxel
X-ray
X-ray crystallography

Information Resource

An element of computing infrastructure (e.g., equipment, personnel) that provides users with the data, information, or knowledge required to help them do their job. Specific information resources may be accessible via the Internet or stored locally on servers.

Cochrane Database
Cumulative Index to Nursing and Allied Health Literature (CINAHL)
Digital library
E-book
Electronic textbook (eBook)
EMBASE
Evidence-based medicine database
Frequently asked questions (FAQ)
Full text database
Genomics database
Guidelines.gov clearinghouse
Hospital Consumer Assessment Healthcare Providers and Systems (HCAHPS)
Impact factor
Index Medicus
Information resources
Internet archive
Medical literature analysis and retrieval system (MEDLARS)
Medicare Provider Analysis and Review (MEDPAR) File
Medicare Provider Inventory (MPI)
Merck Medicus
Multimedia content
Multiparameter Intelligent Monitoring in Intensive Care (MIMIC)
National Death Index (NDI)
National Digital Information Infrastructure Preservation Program (NDIIPP)
National Health Interview Survey (NHIS)
Original content
Physician's Desk Reference (PDR)
Population-based atlas
Primary knowledge-based information
Primary literature
Science citation index
Social security death index
State Medicaid databases

Surveillance, Epidemiology, and End Results (SEER) database
UMLS semantic network
Up-to-date
Value Set Authority Center
Visible human project

Information Retrieval

The process of obtaining information resources (e.g., articles, books, websites) relevant to an information need (i.e., a query) from a collection of information resources. Searches can be based on full-text information resources or other content-based indexing techniques. Often a query does not uniquely identify a single resource from the collection of resources searched, in which case multiple resources are returned and ranked according to different degrees of relevancy (e.g., closeness of match, time of creation, or proximity to the user). This ranking of search results is a key difference of information retrieval searching compared to precise database searching.

All-Payer Claims Database (APCD)
Automated indexing
Bibliographic content
Bibliographic database
Boolean search
Browsing
Check tag
Chronology
Citation database
Cooccurrence of terms
Document frequency
Emtree
Entrez
Entry term
Exact match searching
Excerpta Medica
Exploded term
Field qualification
Filter (for data/information)
Full text
Google
Granularity
Index
Index attribute
Index item
Indexed Sequential Access Method (ISAM)
Indexing
Information
Information need
Information seeking behavior
Inverse document frequency (IDF)

Inverted index
Keyword
Lexical-statistical retrieval
Link-based indexing
Manual indexing
Mesh subheading
Metacontent
Metadata harvester
Natural language query
Online bibliographic searching
Page rank algorithm
Page rank indexing
Partial match searching
Precision
Proximity searching
Publication type
Query
Query and retrieval
Ranking
Recall
Recency ranking
Reference
Relative recall
Relevance ranking
Retrieval
Search intermediary
Search optimization
Set-based searching
Start with versus contains queries
Subheading
Subject heading
Synoptic content
Term frequency
Term frequency—inverse document frequency (TF-IDF)
Term weighting
Text retrieval conference (TREC)
Text word searching
Vector space model
Weights
Wildcard character

Journal

A serious, scholarly, peer-reviewed publication that deals with a particular subject or professional activity. Journals may be available in either paper or electronic formats.

Applied Clinical Informatics (ACI)
Artificial Intelligence in Medicine (AIM)
BMC Medical Informatics and Decision Making
Computers in Biology and Medicine
International Journal of Medical Informatics (IJMI)
Journal of Biomedical Informatics (JBI)
Journal of Clinical Monitoring and Computing
Journal of Medical Internet Research (JMIR)
Journal of the American Medical Association (JAMA)
Journal of the American Medical Informatics Association (JAMIA)
Medical Decision Making
Methods of Information in Medicine (MIM)
Morbidity and Mortality Weekly Report (MMWR)
New England Journal of Medicine (NEJM)

Law

The system of rules that our society recognizes as regulating the actions of its members. Failure to follow these laws may result in the imposition of penalties. Many of the "laws" consist of thousand page–plus documents that describe in excruciating detail what is and what is not allowed to occur. Finally, many of these laws have a significant impact on the way health information technology is or is not used in the clinical setting.

508 compliance
Age Discrimination in Employment Act (ADEA)
Americans with Disabilities Act (ADA)
Antitrust
Any Willing Provider Laws
Arbitration Agreements
Authorized Testing and Certification Bodies (ATCBs)
Belmont Report
Beneficence
Business Associate (BA)
Business Associate Agreement (BAA)
Certified Electronic Health Record Technology (CEHRT)
Clinical Laboratory Improvements Amendment (CLIA)
Code of Federal Regulations (CFR)
Collateral Damages
Conditions of Participation (COP)
Contracting
Copyleft
Copyright
Covered entity
Data stewardship
Data Use Agreement (DUA)
Electronic Protected Health Information (ePHI)
Eligible Hospitals (EH)
Eligible Professionals (EPs)
Emancipated minor
Emergency Medical Treatment and Active Labor Act (EMTALA)
Employee Retirement Income Security Act (ERISA)
End User License Agreement (EULA)
Enterprise Liability
Equal Employment Opportunity (EEO)
Estimated Liability Costs
Fair Information Practice Principles (FIPP)
Family and Medical Leave Act (FMLA)
First do no harm

Freedom of Information Act (FOIA)
Full disclosure
Generalizable knowledge
Guardian
Health Information Technology for Economic and Clinical Health (HITECH) Act
Health Insurance Flexibility and Accountability (HIFA)
Health Insurance Portability and Accountability Act (HIPAA), 1996
Hearsay evidence
Hill–Burton Act
Hold harmless clause
Institutional Review Board (IRB)
Instructional Health Care Directive
Intellectual property (IP)
Justice
Legal issues
Liability
License/Licensure
Licensing
Malpractice
Mandate
Meaningful Use (MU)
Medical power of attorney
Medicare Access and CHIP Reauthorization Act of 2015
Memorandum of Understanding (MOU)
Minor
Moral hazard
Negligence law
Non-Disclosure Agreement (NDA)
Noneconomic Damages
Notifiable disease
Older Americans Act (OAA)
Omnibus Budget Reconciliation Act (OBRA) of 1993
Open access
Open policy
Open source
Patents
Patient Protection and Affordable Care Act (ACA)
Patient Self-Determination Act (PSDA)
Peer review
Plagiarism
Power of Attorney (POA)
Professional patient relationship
Professional review approach

Public Health Service Act
Punitive Damages
Redact
Release of Information (ROI)
Settlement
Software contract
Strict product liability
Subpoena
Tort Reform
Treatment, Payment, Operations (TPO)
View, Download, and Transmit

Logic

The systematic use of symbolic and mathematical techniques and principles underlying the arrangements of elements in a computer, which are used to determine the forms of valid deductive argument with a goal of performing a specific task.

AND (Boolean)
Boolean logic
Complement (Boolean)
If–then–else
Mutually exclusive
NAND (Boolean)
NOT (Boolean)
Not OR (NOR) Boolean logic
OR (Boolean)
Reify
Truth tables
Venn diagram
XOR Exclusive OR (Boolean)

Malware

A relatively new term that is short for "malicious software." Rather than being useful to help solve a problem, this software has a malicious intent to disrupt, damage, steal sensitive information, display unwanted advertising, or disable individual computers or entire computing systems. Malware is not used to describe software that causes unintentional harm due to some error or deficiency in its design, development, configuration, or use.

Adware
Antispyware software
Antivirus software
Bit torrent
Macro virus
Malicious applet
Malicious code
Malicious logic
Malware signature
Phishing
Ransomware
Scareware
Spyware
Trojan horse
Virus (computer code)
Worm
Zombie

Management

The features, functions, and tools required to organize and coordinate the activities of healthcare-related activities within the healthcare system to achieve defined objectives. Management is also concerned with creating the policies and procedures required to organize, plan, control, and direct an organization's resources, including physical, financial, and human to achieve the objectives of the organization (i.e., provide high-quality patient care to the most patients at the least cost).

Accounts payable (AP)
Accounts receivable (AR)
Accreditation
Adult care home
Adult day care
Assessing clinical information system needs
Behavior change
Boiler plate text
Bring your own device (BYOD)
Capital budget
Clinical Decision Support (CDS) Oversight committee
Clinical program
Collaboration
Collaborative decision-making
Collaborative work
Comprehensive Primary Care Initiative
Conflict management
Consensus group process
Consumer price index (CPI)
Cooperative
Credentialing
Data governance
Decision support system (DSS)
Doing Business As (DBA)
Drug Utilization Review (DUR)
Dual reporting structure
Email etiquette
Employee Identification Number (EIN)
Employer Identification Number (EIN)
EMR Oversight committee
Environmental scan
Federally Qualified Health Center (FQHC)
Financial management
Fiscal year (FY)
Fixed cost

Formative decision
Full-time equivalent (FTE)
Global budgeting
Gross domestic product (GDP)
Gross national product (GNP)
Group purchasing organization (GPO)
Health planning
Health Service Area
Horizontal integration
Hospital administration
Incremental cost–benefit ratio
Indefinite Duration, Indefinite Quantity (IDIQ)
Individual practice association (IPA)
Information technology strategy
Initial public offering (IPO)
Inventory
Just-in-time learning
Kaizen
Learning health system
Letter of intent (LOI)
Long run cost
Management by objectives (MBO)
Matrix management
Medical Executive Committee (MEC)
Medical Operations Committee (MOC)
Medical record committee
Medical Review Board (MRB)
Negotiation
Net present value (NPV)
No margin, no mission
Nominal group process
Occupancy rate
On-the-job training (OJT)
Operating budget
Opportunity cost
Opt-in
Opt-out
Organizational behavior
Organizational change
Organizational culture
Organizational mission
Organizational objectives
Organizational tactics
Organizational vision
Participatory decision-making

Patient engagement strategy
Patient safety strategy
Pay by transaction versus pay by user
Performance management
Physician Incentive Plan (PIP)
Population management
Present value
Process improvement
Production
Production room
Productivity cost
Productivity Improvement Program (PIP)
Profit and loss (P&L)
Project management
Project milestones
Prospective payment system
Purchasing collaborative
Reengineering health care
Regional Extension Center (REC)
Request for information (RFI)
Request for proposal (RFP)
Resource allocation
Resources
Responsibility versus accountability
Return on equity (ROE)
Return on investment (ROI)
Risk Management Program
Scope creep
Service Level Agreement (SLA)
Short run cost
Social influence theory
Social media strategy
Software Oversight Committee (SOC)
System development life cycle
System requirements
Time trade-off
Utilization review
Value-added reseller (VAR)
Value-based purchasing (VBP)
Volume performance standard
Work to limit of license
Workflow reengineering
Zero-base budgeting (ZBB)

Mathematics

An abstract field of science concerned with the study of topics such as number, quantity, structure, patterns, space, and change of physical objects and abstract concepts. Those who study mathematics (mathematicians) look for patterns and attempt to use them to formulate conjectures. They then attempt to resolve the truth or falsehood of the conjectures by mathematical proof. When mathematicians create mathematical structures that are good models of real phenomena, then mathematical reasoning can provide insight or predictions about nature or the future.

2.71828 (e)
3.14159 (pi)
1024 (2^10)
Acute
Arc
Arithmetic moving average (ARIMA)
Base-10
Base-2
Billion
Binary
Binary coded decimal (BCD)
Binary to decimal conversion
Boolean operators
Cardinal numbers
Coefficient
Complex numbers
Constant
Cosign vector calculation
Cosine function
Cross product
Date
Default value
Denominator
Denormalized numbers
Density–amplitude domain
Derivative
Differential
Differential equation
Differentiate
Digital
Dimension
Discrete
Discriminant

Dot product
Eigenvectors
Entropy
Exponent
Exponential constant
Exponential function
Factor
Factorial
Floating-point arithmetic
Fourier analysis
Frequency components
Frequency response
Function
Fuzzy logic
Global extrema
Gradient
Gradient descent
Halting problem
Hash value
Hyperbolic
Infinitesimal
Infinity
Integral
Interpolate
Intersection
Inverse
Inverse function
Kilobyte
Limit
Linear
Linear algebra
Local extrema
Logarithm
Matrix
Matrix inversion
Maximum
Multiplication
Natural logarithm
Nonlinearity
Not a number (NaN)
Numerator
Octal
Operator
Optimization
Order of magnitude

Order of operations
Ordinal numbers
Partial derivative
Percentage
Perimeter
Power
Power law
Proof
Proportional
Quadratic equation
Quadratic formula
Quotient
Ratio
Reciprocal
Regression
Residual
Root
Root mean square
Satisfiability problem
Sign
Sine function
Smoothing
Space constant
Sum of squares
Summation
Tangent
Time constant
Travelling salesman problem
Union
Vector differentiation
Wavelet transformation

Measurement

The assignment of a number to some characteristic of an object or event (e.g., size, length, amount, time, temperature) so that it can be compared with other objects or events. The scope and application of a measurement is dependent on the context and discipline. In the natural sciences and engineering, measurements do not apply to nominal properties (i.e., properties with no magnitude) of objects or events. However, in other fields such as statistics as well as the social and behavioral sciences, measurements can have different levels, including nominal, ordinal, interval, and ratio scales.

Accuracy of measurement
Acoustic
Acuity
Analog
Analytical
Angstrom
Attenuation
Calibration
Calorie
Distortion
Drift
Dye dilution indicator
Electromyogram
False negative result
False positive result
Fractional change
Gain
Gold standard test
Index test
Indicator
Intermittant monitoring
Invasive monitoring
Key performance indicator (KPI)
Off scale
Offset
Order entry rate
Patient satisfaction
Point of service (POS) testing
Pressure
Pressure transducer
Quantity
Real-time data acquisition
Recording

Refractivity index
Resolution of measurement
Response time
Sampling rate
Scalar
Signal
Signal artifacts
Spectrum bias
Static
Temporal
Transducer
Ultrasonography
Unobtrusive measures
Valid
Visual analog scale
Within Defined Limits (WDL)

Measurement Unit

A term used to describe a definite magnitude of a quantity, defined and adopted by convention or by law. In general, a country or organization adopts a specific set of measurement units which are used as a standard for measurement of the same quantity everywhere. There are several different overarching measurement systems including the "English" system, which is in use in the United States, and the SI (Système International d'Unités), which is a globally agreed upon system of units, with seven base units. These base units can be modified by various prefixes (e.g., milli, kilo, mega, etc.). Using the SI system, any value of a specific quantity can be expressed as a simple multiple of the unit of measurement. For most scientific purposes, the SI measurement units are used regardless of the country in which one resides.

Ampere (A)
Atto (a)
Candela (cd)
Celsius (C)
Centigrade (C)
Cubic centimeters (cc)
Decibel (Db)
Exa (E)
Fahrenheit (F)
Femto (f)
Giga (G)
Gram (g)
Hertz (Hz)
Hour (H)
Inch (In)
Kelvin (K)
Kilo (k)
Kilogram (kg)
Mega (M)
Meter (m)
Metric
Micro (μ)
Microgram (mcg)
Micron
Milli (m)
Milliequivalent (meq)
Milligram (mg)
Milliliter (mL)
Millimeter (mm)
Mole (mol)

Nano (n)
Newton
Parts per million (PPM)
Peta (P)
Pico (p)
Postmeridiem (PM)
Radian
Revolutions per minute (RPM)
Second (s)
Tera (T)
Tesla
Units
Yocto (y)
Yotta (Y)
Zepto (z)
Zetta (Z)

Medical Billing

The process by which medical coders (either automated or manually) translate clinical documentation of a healthcare service, which is often described in narrative or free text, into a set of diagnosis, procedure, or medication codes, which can be submitted as a claim to an insurance company or directly to the patient. Often, the healthcare provider must follow-up on the claim with the insurance company to ensure it is paid.

Administrative services only (ASO)
Admitted Carriers
All Patient Diagnosis-Related Group (APDRG)
Allowable Costs
All-Payer System
Alternative Market
Ambulatory Payment Classification (APC)
Appropriateness
Assisted Living Facility (ALF) Benefit Amount
Average Wholesale Price (AWP) of Prescription Drugs
Avoidable Hospital Conditions
Bad Debts
Balance Billing
Basis of Eligibility (BOE)
Beneficiary
Benefit Start Date of Current Claim Period
Billing Audit
Cafeteria Benefits Plan
Calendar Year
Capitated System
Capitation
Categorically Needy
Charges
Charity Care
Chart Audit
Claim Status
Coinsurance
Company Code
Contract Management System
Coordination of Benefits (COB)
Copayment
Cost center
Customary Charge
Customary, Prevailing, and Reasonable
Customer
Deductible

Diagnosis-Related Group (DRG)
Direct cost
Discounting
Disease Management Program
Durable Medical Equipment (DME)
Electronic Claim
Electronic Funds Transfer (EFT)
Employer Master File (EMP)
Employer Name
Employer Type
Evaluation and Management Codes (E&M)
Explanation of Benefits (EOB)
Federal Employer Identification Number (FEIN)
Fee for Service (FFS)
Fee Schedule
Fiduciary
Fixed fee
Foster Child
Fraud
Hospital Acquired Condition (HAC)
Indigent Care
Indirect cost
Inpatient Prospective Payment System
Level of Care (LOC)
Marginal cost
Medical Record Number (MRN)
Medical Savings Account (MSA)
Medically Indigent
Merit-Based Incentive Payment System
National Plan and Provider Enumeration System
National Provider Index (NPI)
Nonquantifiable benefits and costs
Orderable, Performable, Chargeable (OPC)
Overhead
Per capita payment
Pharmacy Benefit Managers (PBMS)
Physician Fee Schedule
Preferred Provider Insurance
Prepayment
Present on Admission (POA)
Primary Care Gatekeepers
Prospective payment
Relative Value Unit (RVU)
Resource-Based Relative Value Scale
Retrospective payment

Revenue center
Service benefit
Service bureau
Stop loss coverage
Super bill
Taxpayer Identification Number (TIN)
Universal Billing Form 92 (UB-92)
Usual customary and reasonable fee
Variable cost
Willingness to pay

Medical Device

The U.S. Food and Drug Administration (FDA) defines a medical device as an instrument, apparatus, implement, machine, contrivance, implant, in vitro reagent, or other similar or related article, including a component part, or accessory that is intended for use in the diagnosis of disease or other conditions, or in the cure, mitigation, treatment, or prevention of disease, in man or other animals, or intended to affect the structure or any function of the body of man or other animals, and which does not achieve any of its primary intended purposes through chemical action within or on the body of man or other animals and which is not dependent upon being metabolized for the achievement of any of its primary intended purposes.

Assistive devices
Automatic inflation protection type
Cannula
Cardiovascular monitor
Catheter
Clamp
Crash cart
Defibrillator
Electrocardiogram (ECG/EKG)
Heart–Lung pump
High-efficiency particulate attraction filter (HEPA)
Hoist scale
Home medical equipment
Implantable cardioverter defibrillator (ICD)
Intraortic balloon pump (IABP)
Intravenous (IV) pump
Laminar airflow hood
Left ventricular assist device (LVAD)
Ligature
Medication cart
Nasal cannula
Nasogastric tube (NG)
Patient-controlled analgesia (PCA) pump
Percutaneous endoscopic gastrostomy (PEG) tube
Peripherally inserted central catheter (PICC)
Positive end-expiratory pressure (PEEP)
Pyxis machine
Radio-frequency identification device (RFID)
Ventilator
Wheelchair (WC)

Medical Facility

A physical location where health care is provided. Medical facilities can range from small clinics and single physician offices to medium-sized urgent care centers and large hospitals with elaborate emergency rooms and trauma centers. In most places, medical facilities are regulated to some extent by a government or private entity. Such licensing by an approved regulatory agency is often required before a facility may open for business and care for patients. Medical facilities may be owned and operated as or by for-profit businesses, nonprofit organizations, and local, state, or federal governments.

Academic Medical Center (AMC)
Acute Care Unit
Ambulatory Care
Ambulatory Clinic
Ambulatory Surgical Center (ASC)
Area Health Education Center (AHEC)
Assisted Living Facility
Behavioral Health
Board and Care Home
Children's Hospital
Chronic Care
Clinic
Community Health Center (CHC)
Community Hospital
Community Long-Term Care (CLTC)
Community Mental Health Center (CMHC)
Community-Based Care/Services
Comprehensive Cancer Center
Continuing Care Retirement Community (CCRC)
Critical Access Hospital (CAH)
Emergency Care Center
Emergency Medical Services (EMS)
Emergency Shelter
Family Foster Home
Foster Care
Free clinic
General Practice
Geriatric Research, Education, and Clinical Center (GRECC)
Group Home
Group Practice
Health Facilities
Home Health

Home Health Care
Hospice
Hospital
Independent Living Facility
Intermediate Care Facility (ICF)
Intermediate Care Facility for the Mentally Retarded (ICF/MR)
Isolation room
Long-Term Care (LTC)
Medicare HMOs
Memory Care Unit
Military Treatment Facilities (MTFs)
Mobile Army Surgical Hospital (MASH)
Neighborhood Health Center
Nursing Home
Nursing Home Care
Outpatient
Private Practice
Rehabilitation Hospital
Residential Care
Secure Facility
Senior Center
Skilled Nursing Facility (SNF)
Step down unit
Tertiary Care Hospital
Transitional Care
Urgent Care Center (Clinic)
Veteran Integrated Service Networks (VISN)
Wellness Clinic

Medication

A chemical substance that is introduced (i.e., ingested, injected into a muscle or vein, applied topically to the skin, inhaled, or inserted rectally) into a patient's body. A medication is designed to treat a patient's physical or mental illness or to relieve one or more symptoms of a patient's clinical condition. It is not uncommon for medications to have unintended, adverse effects, or to interact with each other and harm patients. Often medications are referred to as drugs or pharmaceuticals.

Adrenaline
Albuterol (Proventil)
Alendronate (Fosamax)
Allopurinol (Zyloric)
Alprazolam (Xanax)
Amitriptyline (Elavil)
Amlodipine (Norvasc)
Amoxicillin (Trimox)
Amoxicillin–clavulanate (Augmentin)
Analgesia
Antibiotic (ABX)
Artificial nutrition and hydration
As desired (ad lib)
As needed (prn)
Aspirin (asa)
At bedtime (q hs)
Atenolol (Tenormin)
Atomoxetine (Strattera)
Atorvastatin (Lipitor)
Azidothymidine (AZT)
Azithromycin (Zithromax)
Azithromycin pack (Z-pack)
Bedtime (hs)
BID—twice daily
Biocompatible
Biosimilar
Bupropion (Wellbutrin)
Capsule (cap)
Celecoxib (Celebrex)
Cetirizine (Zyrtec)
Ciprofloxacin (Cipro)
Citalopram (Celexa)
Clonazepam (Klonopin)
Clonidine (Catapress)
Clopidogrel (Plavix)

Interchangables
International units (IU)
Intravenous (IV)
Iron supplements
IV Piggyback (ivpb)
IV Push (ivp)
Lansoprazole (Prevacid)
Levofloxacin (Levaquin)
Levothyroxine, T4 (Synthroid)
Lipids
Lisinopril (Prinivil)
Lorazepam (Ativan)
Losartan (Cozaar)
Losartan–hydrochlorothiazide (Hyzaar)
Lovastatin
Magnesium sulfate ($MgSO_4$)
Measles, Mumps, Rubella (MMR) vaccine
Medication name
Medication substitution
Metformin (Glucophage)
Methylphenidate (Ritalin)
Methylprednisolone (Medrol)
Metoprolol
Morphine (MSO_4)
Naproxen (Naprosyn)
Narcan
Niacin (Niacor)
Nifedipine (Procardia)
Nitroglycerin (Nitrolingual)
Normal Saline (ns)
Nothing per os (NPO)
Number of doses
Olanzapine (Zyprexa)
Omeprazole (Prilosec)
Once daily (OD)
Orally (Per os)
Over the counter (OTC)
Oxycodone (OxyContin)
Oxygen (O_2)
Pantoprazole (Protonix)
Paroxetine (Paxil)
Patient-controlled analgesia (PCA)
Penicillin V (Veetids)

Phenytoin (Dilantin)
Pioglitazone (Actos)
Pitocin (Oxytocin)
Pneumococcal conjugate vaccine
Potassium chloride (Klor-Con)
Pravastatin (Pravachol)
Prednisone (Deltasone)
Prescription (Rx)
PRN—from the Latin "pro re nata"
Promethazine (Phenergan)
Propranolol (Inderal)
Proton pump inhibitors (PPI)
Quetiapine (Seroquel)
Quinapril (Accupril)
Ramipril (Altace)
Ranitidine (Zantac)
Ringer's solution
Risperidone (Risperdal)
Rosiglitazone (Avandia)
Rx—perscription
Saline
Sertraline (Zoloft)
Sildenafil (Viagra)
Simvastatin (Zocor)
Sodium chloride (NaCl)
Spironolactone (Aldactone)
Subcutaneous (sq)
Sulfamethoxazole–trimethoprim (Bactrim)
Three times a day (tid)
To keep open (TKO)
Tobramycin–dexamethasone (TobraDex)
Topiramate (Topamax)
Total parenteral nutrition (TPN)
Toxin
Tramadol (Ultram)
Triamterene–hydrochlorothiazide (Dayside)
Twice daily (bid)
Unit-based dosing
Valsartan (Diovan)
Verapamil (Isoptin)
Warfarin (Coumadin)
Zolpidem (Ambien)

Natural Language Processing

A subfield of computer science, artificial intelligence, and computational linguistics concerned with developing algorithmic techniques to enable computers to understand human-generated written or spoken natural language.

Acronym expansion
Acronym standardization
Anaphoric references
Annotate
Annotation
Bigrams
Bound morpheme
Clauses
Concept hierarchy
Context deficit
Context-free grammar
Contextual meaning
Corpora
Derivational morpheme
Double negative detection
Exception rules
Exceptions
Foreign language detection
Grammar
Inflectional morpheme
Information extraction
Irony detection
Label
Language modeling
Lexeme
Lexical form
Lexical variants
Lexicon
Logical connections
Machine translation
Map tables
Metamap
Morpheme
Natural language processing (NLP)
Negation
Negative dictionary
N-grams
Parse

Parse tree
Part of speech tagging
Probabilistic context free grammar
Profanity detection
Punctuation
Punctuation correction
Referential expression
Sarcasm detection
Semantic analysis
Semantic grammar
Semantic mapping
Semantic pattern
Semantic type
Sentiment analysis
Specialized vocabulary mapping
Spelling check
Spelling correction
Stem
Stop word list
Stop words
String
Syntax verification
Term
Text parsing
Tokenization
Tokens
Trigrams
Valence weighting
Vector mapping
White space
Word duplication
Word sense
Word sense disambiguation

Network Security

The activities, policies, and practices an organization uses to protect, prevent, and monitor for unauthorized access, misuse, unintentional modification, or denial of access to a computer network and the network-accessible data, information, and knowledge it contains. Network security involves the authorization of access to data in a network, which is controlled by the network administrator. It includes both hardware and software technologies.

Active attack
Airgap
Asymmetric cryptography
Attack
Attack method
Attack mode
Attack pattern
Attack signature
Attacker
Back door
Blacklist
Bot
Bot herder
Bot master
Botnet
Computer network defense analysis
Critical infrastructure
Cyber infrastructure
Cybersecurity
Denial of service
Distributed denial of service
Drive-by download
Dynamic attack surface
Firewall
Hacker
Intrusion
Intrusion detection
IP sec (Internet Protocol Security) logic bomb
Network mapper (Nmap)
Network resilience
Next-generation firewall
Packet capture
Penetration
Penetration test (Pen test)
Perimeter definition

Pretty Good Privacy (PGP)
Private key
Proxy access
Public key
Public key cryptography
Public key encryption
Public key infrastructure (PKI)
Remote desktop protocol (RDP)
Secure Socket Layer (SSL)
System integrity
System security analysis
System security architecture
Virtual private networks (VPNs)

Organization

A legal entity consisting of an organized group of people that has a collective purpose or goal. There are a variety of types of organizations, including business corporations, governments, nongovernmental organizations, political organizations, international organizations, armed forces, charities, societies, associations, not-for-profit corporations, partnerships, cooperatives, and educational institutions.

Accountable Care Organization (ACO)
Ambulatory Care Group (ACG)
Cochrane Collaboration
Community Health Information Network (CHIN)
Community Health Management Information Systems (CHMIS)
Digital Preservation Coalition
Disease Management
For-Profit
Foundation for Accountability (FACCT)
Group model health maintenance organization
Health care organization (HCO)
Health Maintenance Organization (HMO)
Home and Community-Based Services (HCBS)
Institutional Health Services
Internet Service Provider (ISP)
Local Service Provider
Managed Care Organization (MCO)
Managed competition
N3 Service Provider (N3SP)
National Application Service Provider (NASP)
Nationwide Health Information Network (NwHIN)
Network model HMO
Network-model health maintenance organization
Nonprofit/Not-For-Profit
Open archives initiative
Pharmaceutical Benefits Manager (PBM)
Physician Hospital organization
Preferred Provider Organization (PPO)
Prepaid group practice
Provider Sponsored Organization (PSO)
Purchasing coalitions
Regional network
Social Health Maintenance Organization (SHMO)
Staff model health maintenance organization (HMO)
The Joint Commission (TJC)
The Joint Commission for Accreditation of Healthcare Organizations (JCAHO)
World intellectual property organization

Patient Safety

The field of study that focuses on the policies, procedures, and activities designed to prevent accidental or preventable harm (e.g., medical errors, injuries, accidents, and infections) produced in the course of providing medical care to patients. Research and activities emphasize the system of care delivery rather than focusing entirely on the activities of individuals that are designed to prevent errors, and to learn from errors that occur. The field is built on the concept of a blame-free culture of safety that involves healthcare professionals, organizations, and patients.

Active error
Adverse drug event (ADE)
Adverse drug reaction (ADR)
Adverse event
Blame-free culture
Blunt end
Checklist
Checklist effect
Clinical risk
Clinical risk analysis
Clinical risk control
Clinical risk estimation
Clinical risk evaluation
Clinical risk management
Clinical risk management file
Clinical risk management plan
Clinical risk management process
Clinical safety
Clinical safety case
Clinical safety case report
Clinical safety officer
Close call
Commercial off-the-shelf (COTS) product
Common format
Failure mode and effects analysis (FMEA)
Fixation error
Harm
Hazard
Hazard log

Health IT System
Hospital-acquired
Iatrogenic event
Incident
Initial clinical risk
Intended use
Issue
Likelihood
Manufacturer
Never events
Nonhealth product
Nosocomial
Patient
Postdeployment
Procedure
Release
Residual clinical risk
Risk
Risk factor
Safety incident
Safety incident management log
Sentinel event (SE)
Sentinel event alert (SEA)
Serious safety event (SSE)
Severity
Technology-related event (TRE)
Top management
Unintended consequences

People

The men and women who have made important contributions to the field of clinical informatics. This category also includes the titles or roles within the organization that people routinely fill.

Al Barrak, Ahmed
Al-Shorbaji, Najeeb
Altman, Russ
Altuwaijri, Majid
Ameen, Abu-Hanna
Ammenwerth, Elske
Andersen, Stig Kjaer
Aronsky, Dominik
Bakken, Suzanne
Bakker, Ab
Ball, Marion J.
Barnett, G. Octo
Bates, David W.
Bellazzi, Riccardo
Bleich, Howard L.
Blobel, Bernd
Borycki, Elizabeth
Brennan, Patricia Flatley
Butte, Atul
Carr, Robyn
Chang, Polun
Chief Clinical Informatics (Information) Officer (CCIO)
Chief Privacy Officer
Chute, Christopher G.
Cimino, James J.
Classen, David C.
Coiera, Enrico
Collen, Morris F.
Curmudgeon
Database Administrator (DBA)
De Moor, Georges
Degoulet, Patrice
Detmer, Don E.
Engelbrecht, Rolf
Espinosa, Amado
Eysenbach, Gunther
Fieschi, Marius
Fox, John
Friedman, Carol
Friedman, Charles P.

Luna, Daniel
Mandil, Salah Hussein
Mandl, Ken D.
Mantas, John
Maojo, Victor
Marcelo, Alvin
Margolis, Alvaro
Marin, Heimar de Fatima
Martin-Sanchez, Fernando
Masic, Izet
McCray, Alexa
McDonald, Clement "Clem" J.
Mihalas, George
Miller, Perry L.
Miller, Randolph A.
Moehr, Jochen
Moen, Anne
Moghaddam, Ramin
Moore, Jason H.
Moura, Lincoln de Assis
Murray, Peter
Musen, Mark A.
Nohr, Christian
Norman, Donald A.
Ohno-Machado, Lucila
Otero, Paula
Park, Hyeoun-Ae
Patel, Vimla L.
Peterson, Hans
Pinciroli, Francesco
Protti, Denis
Rector, Alan
Rienhoff, Otto
Ritchie, Marylyn D.
Roberts, Jean
Roger France, Francis
Sabbatini, Ranato
Safran, Charles
Saltz, Joel H.
Saranto, Kaija
Scherrer, Jean-Raoul
Schneider, Werner
Seroussi, Brigitte
Shabo, Amnon

Shahar, Yuval
Shaikh, Aziz
Shortliffe, Edward "Ted" H.
Slack, Warner V.
Smith, Barry
Stead, William W.
Szolovits, Peter
Takeda, Hiroshi
Talmon, Jan
Tanaka, Hiroshi
Tchuitcheu, Ghislain Kouematchoua
Tierney, William M.
Toyoda, Ken
van Bemmel, Jan H.
van der Lei, Johan
Warner, Homer R.
Weber, Patrick
Weed, Lawrence
Westbrook, Johanna
Westbrooke, Lucy
Wiederhold, Giovanni "Gio" C. M.
Wong, Chun-Por (CP)
Wright, Graham
Wu (Ying Wu), Helen
Wyatt, Jeremy
Zhao, Junping
Zvarova, Jana

Physiologic Measurement

Techniques used to assess or measure the function of major organ systems or other bodily functions either directly or indirectly. Physiological measurements can be obtained using a variety of methods, including: self-report; direct observation; direct measurement; indirect measurement; laboratory tests; and electronic monitoring.

ABO blood group (ABO)
Acid–base balance
Acidosis
Activated partial thromboplastin time—aPTT (PTT)
Acute physiology and chronic health evaluation (APACHE)
Acute physiology score (APS)
Acute renal failure (ARF)
Alkalosis
Alveolar to arterial partial pressure of oxygen gradient
(A–a gradient)
Amylase
Antinuclear antibodies (ANA)
Apgar score
Basic metabolic panel (BMP)
Birth weight (BW)
Blood pressure (BP)
Blood urea nitrogen (BUN)
Body mass index (BMI)
Body surface area (BSA)
Complete blood count (CBC)
Comprehensive metabolic panel (CMP)
Creatine clearance (CrCl)
Diastole
Draw time
Eindhoven's triangle
Electroencephalogram (EEG)
Electrolyte balance
Electrolyte panel (lytes)
Estimated creatinine clearance
Fluid balance
Glasgow Coma Score (GCS)
Glomerular filtration rate (GFR)
Health evaluation (Apache–II) scoring system
Height (Ht)
Hematocrit (hct)
Hemoglobin (Hb)
Hemoglobin A1C or Glycohemoglobin (HbA1C)

Physiology

A subfield of biology that deals with the normal functions of living organisms and their parts. Physiologists focus on how organisms, organ systems, organs, cells, and biomolecules carry out the chemical or physical functions that are required to maintain a living system.

Absolute refractory period
Absorption
Acclimation
Accommodation
Action potential
Activation
Active transport
Adaptation
Aerobic
Alimentary
Ambient
Amplification
Anoxia
Antagonist
Arteriovenous
Balance
Bowel movement (BM)
Catabolism
Circadian rhythm
Compartment
Compliance
Contraction
Dead on arrival (DOA)
Differentiation
Diffusion

Digestion
Dose response curve
Equilibrium
Evoked
Excretion
Exsanguinate
Fibrillation
Gustatory
Habituation
Homeostasis
Ingestion
Inhibition
Leak
Live
Long-term memory
Mechanism
Mental status
Metabolism
Metabolite
Micturition
Motility
Phase
Plasticity
Prandial
Receptor
Recurrent
Reflex
Refractory period
Regulate
Respiration
Retro grade
Secretion
Short-term memory
Steady-state
Stimulus
Stimulus response
Supramaximal
Systole
Threshold
Transport
Upregulation
Ventilate
Ventilation

Probability Distribution

A mathematical description of a particular phenomenon in terms of the probabilities of events. Examples of such phenomena include the measurement of naturally or man-made events, the results of an experiment, or a survey. A probability distribution is defined in terms of an underlying sample space, which is the set of all possible outcomes of the phenomenon being observed. The sample space may be the set of real numbers or a higher-dimensional vector space, or it may be a list of nonnumerical values (e.g., the sample space of a coin flip would be heads or tails).

Bernoulli distribution
Bimodal distribution
Binomial distribution
Exponential distribution
Gaussian distribution
Kurtosis
Log-normal distribution
Normal distribution
Poisson distribution
Power law distribution
Skewed
Skewness
Uniform distribution

Professional Organization

Most often a nonprofit organization with the goal of furthering the mission of a particular profession, maintaining control, or oversight of the legitimate practice of those in the profession and their privileged and powerful position as a controlling body, promoting the interests of those individuals engaged in that profession, and safeguarding the public's interest in the field. Many professional organizations are involved in the development and monitoring of professional and academic educational programs, and updating the skills of its membership. Often the organization is responsible for overseeing or promoting professional certification within their field to indicate that a person possesses the required qualifications to practice safely and effectively in their specific subject area. Finally, many professional organizations act as learned societies for the academic disciplines underlying their professions.

American Academy of Family Physicians (AAFP)
American Academy of Pediatrics (AAP)
American Cancer Society (ACS)
American Civil Liberties Union (ACLU)
American College of Medical Informatics (ACMI)
American College of Radiology (ACR)
American Dental Association (ADA)
American Diabetes Association (ADA)
American Health Information Management Association (AHIMA)
American Heart Association (AHA)
American Hospital Association (AHA)
American Lung Association (ALA)
American Medical Association (AMA)
American Medical Informatics Association (AMIA)
American Nurses Association (ANA)
American Psychiatric Association (APA)
American Psychological Association (APA)
American Public Health Association (APHA)
Association for Retarded Citizens (ARC)
Association of Medical Directors of Information Systems (AMDIS)
Canadian Medical Association (CMA)
College of American Pathology (CAP)
College of Healthcare Information Management Executives (CHIME)
Computer-Based Patient Record Institute (CPRI)

Electronic Health Record Association (EHRA)
Health Information Management and Systems Society (HIMSS)
Healthcare Information and Management Systems Society (HIMSS)
Independent Physician Association (IPA)
Independent Practice Association (IPA)
Institute for Healthcare Improvement (IHI)
Institute for Safe Medication Practices (ISMP)
Institute of Electrical and Electronic Engineers (IEEE)
Institute of Medicine (IOM)
International Academy of Health Sciences informatics (IAHSI)
International Medical Informatics Association (IMIA)
Medical Group Management Association (MGMA)
Medinfo
Mothers Against Medical Error (MAME)
National Academy of Engineering (NAE)
National Academy of Medicine (NAM) [formerly, Institute of Medicine (IOM)]
National Academy of Science (NAS)
National Alliance for Health Information Technology (NAHIT)
National eHealth Transition Authority (NeHTA)
National Electrical Manufacturers Association (NEMA)
Object Management Group (OMG)
Professional Standards Review Organization (PSRO)
Radiological Society of North American (RSNA)
Visiting Nurse Association (VNA)
Workgroup (WG)

Programming Language

A formal computer language that includes a controlled vocabulary and set of grammatical rules designed to instruct a computer how to perform specific tasks. Programming languages are used to create programs to control the behavior of a machine or to express algorithms. The description of a programming language is usually split into two components: syntax (form) and semantics (meaning).

Assembly language

C Sharp (C#)

C++

Cache

Common Business-Oriented Language (COBOL)

Fortran—FORmula TRANslation

Hypertext Markup Language (HTML)

Hypertext Preprocessor (PHP)

Java

JavaScript

Job Control Language (JCL)

LISP (LISt Processor)

Machine language

Markup language

Massachusette's General Hospital (MGH) Utility Multi-Programming System (MUMPS)

Mathematical Markup Language (MathML)

MicroArray and Gene Expression Markup Language (MAGE-ML)

Object Constraint Language (OCL)

Ontology Web Language (OWL)

Perl

Python

R programming language

Ruby on Rails

Structured Query Language (SQL)

Swift

Symbolic programming language

Quality Management

A business philosophy, focused on customer satisfaction that leads to a set of actions or system to manage the activities and tasks needed to maintain a desired level of consistency or even excellence within a product, process, service, or business. It has four main components: quality planning, quality assurance, quality control, and quality improvement. Quality management is focused not only on product and service quality, but also on the means to achieve it.

Access and equity for patient populations
Apples-to-apples comparison
Automated measure submission to CMS
Average length of stay (ALOS)
Benchmark
Case mix adjustment
Case Mix Index (CMI)
Case-rate
Catheter-associated urinary tract infection (CAUTI)
Central line-associated bloodstream infection (CLABSI)
Charlson comorbidity index
Clinical performance measures
Clinical quality measure
Continuous quality improvement (CQI)
Cost-effectiveness analysis (CEA)
Customer focus
Define, Measure, Analyze, Improve, Control (DMAIC)
Effective
Efficiency
Efficient
Electronic Clinical Quality Measure (eCQM) (eMeasure)
Engagement of people
Episode of care
Equitable
Evidence-based decision-making
Expanded quality assurance (XQA)
Experience rating
Fraud, waste, and abuse (FWA)
Health Plan Employer Data and Information Set (HEDIS)
Healthcare-acquired infection (HAI)
Healthcare Cost and Utilization Project Quality Indicators (HCUP QIs)
Healthcare Effectiveness Data and Information Set (HEDIS)
Health-related quality of life (HRQL)
Hospital acquired infection (HAI)

Inpatient quality reporting
Instrumental activities of daily living (IADLs)
Leadership
Lean management
Length of stay (LOS)
Level of care criteria
Medical Outcomes Study 36 Item Short Form Health Survey (SF-36)
Metrics
Morbidity
Mortality
National Patient Safety Foundation (NPSF)
National Patient Safety Goal (NPSG)
Pareto principle
Patient safety indicator (PSI)
Patient-centered
Pay for Performance (P4P)
Peer Review Organization (PRO)
Physician Quality Reporting Initiative (PQRI)
Physician Quality Reporting System (PQRS)
Process approach
Producer price index (PPI)
Quality assurance (QA)
Quality control (QC)
Quality improvement (QI)
Quality improvement strategy
Quality management system
Quality measurement (management) dashboard
Quality of Care
Quality planning
Quality Reporting Data Architecture
Rapid-cycle improvement
Relationship management
Reporting period
Root-cause analysis (RCA)
Safe
Severity of illness
Six Sigma
Surgical Quality Alliance (SQA)
System improvement
Timely
Total quality improvement/management (TQI/TQM)
Value of a statistical life (VSL)
Ventilator-associated pneumonia (VAP)
Zero defects (ZD)

Screening Test

Laboratory or radiology tests used to identify individuals within a population who are at an increased risk for a clinical condition (e.g., high cholesterol levels) or disease (e.g., mammogram for breast cancer) before they have signs, symptoms, or even realize they may be at risk so that preventive measures can be taken. They are most valuable when they are used to screen for diseases that are both serious and treatable, so that there is a benefit to detecting the disease before symptoms begin at their most treatable stages. Good screening tests should be highly sensitive, or able to accurately identify those individuals who might have a given disease. A positive screening test often requires further testing with a more specific test or one that is better able to correctly exclude those individuals who do not have the given disease or to confirm a diagnosis.

Autism screening
Behavioral assessments
Blood pressure screening
Body mass index (BMI) measurements
Cervical dysplasia screening
Depression screening
Developmental screening
Dyslipidemia screening
Fluoride chemoprevention supplements
Gonorrhea preventive medication
Hearing screening
Hematocrit or hemoglobin screening
Hemoglobinopathies
Hepatitis B screening
Human immunodeficiency virus (HIV) screening
Human papillomavirus screening test (Pap smear)
Hypothyroidism screening
Lead screening
Obesity screening and counseling
Oral health risk assessment
Phenylketonuria (PKU) screening
Sexually transmitted infection (STI) prevention, counseling, and screening
Sickle cell screening
Tuberculin testing
Vision screening

Standard

A standard, or well-accepted, uniform set of terms, concepts, procedures, structures, or capabilities, that have been carefully defined and agreed upon by a respected organization, is necessary to allow computers to transfer data, information, or knowledge from one device or application to another.

American Standard Code for Information Interchange (ASCII)
Arden syntax
Association
Clinical Context Object Workgroup (CCOW)
Clinical Document Architecture (CDA)
Common data elements
Common Industry Format (CIF)
Conformance Statement
Consolidated Clinical Document Architecture (C-CDA)
Continuity of Care Document (CCD)
Continuity of Care Record (CCR)
Cross-Enterprise Document Sharing (XDS)
Curly braces problem
Data interchange standard
Data standards
De jure standard
Defacto standard
Digital European cordless telephone (DECT)
Digital Imaging and Communications in Medicine (DICOM)
Digital Object Identifier (DOI)
Direct protocol
Domain Name System (DNS)
Draft Standard for Trial Use (DSTU)
Dublin Core Metadata Initiative (DCMI)
Extended Binary Coded Decimal Interchange Code (EBCDIC)
eXtensible Mark-up Language (XML)
Fast Healthcare Interoperability Resources (FHIR)
Federal Information Processing Standards (FIPS)
File Transfer Protocol (FTP)
Formal standard
Graphics Interchange Format (GIF)
Guideline Interchange Format (GLIF)
Hypertext transfer protocol (http)
Hypertext Transfer Protocol Secure (HTTPS)
Infobutton
Integrated Services Digital Network (ISDN)
Integrating the Healthcare Enterprise (IHE)

International Standard Book Number (ISBN)
Internet address (IP address)
Internet Control Message Protocol (ICMP)
Internet Mail Access Protocol (IMAP)
Internet Protocol (IP)
Internet standards
Interoperability standards
Joint Photographic Experts Group (JPEG)
Lightweight Directory Access Protocol (LDAP)
Logical Observation Identifiers Names and Codes (LOINC)
Message
Messaging standards
Multipurpose Internet mail extensions (MIME)
Network Time Protocol (NTP)
Open System Interconnection (OSI)
Patient identifier (unique, national)
Portable Document Format (PDF)
Portable Operating System Interface Exchange (POSIX)
Post office protocol (POP)
Privacy enhanced mail (PEM) protocol
Protocol for metadata harvesting
Reference Information Model (RIM)
Resource description framework
RS-232
Secure file transfer protocol (SFTP)
Secure Multipurpose Internet Mail Extensions (S-MIME)
Security Assertion markup Language (SAML)
Simple mail transport protocol (SMTP)
Standard development process
Standard Generalized Markup Language (SGML)
Standard international (SI) system of units
Standard Protocol and RDF Query Language (SPARQL)
Structured Mark-up Language (SML)
Technology Readiness Levels (TRL)
Transaction standards
Transmission Control Protocol (TCP)
Transmission Control Protocol (TCP) and the Internet Protocol
(IP) (TCP/IP)
Unicode
Unified Medical Language System (UMLS)
XDR and XDM for Direct Messaging specification
XML format
XML Paper Specification (XPS)
Z-segment (HL-7 v2.x)

Standards Organization

A standards organization's [also referred to as a standards body, standards developing organization (SDO), or standards setting organization (SSO)] primary activities include developing, coordinating, promulgating, revising, amending, reissuing, interpreting, or otherwise producing technical standards. The resulting standards are intended to address the needs of a group of affected adopters (e.g., product or service developers, purchasers, and users). Most standards are voluntary in that they are offered for adoption by groups or industry without being mandated in law. Some standards become mandatory when they are adopted by regulators as legal requirements in particular domains.

American National Standards Institute (ANSI)
American Society for Testing and Materials (ASTM)
Clinical Data Interchange Standards Consortium (CDISC)
Conseil European pour la recherche nucleaire (CERN)
European Committee on Standardization (CEN)
Health Informatics Standards Board (HISB)
Health Level Seven (HL-7 or HL7)
Health on the Net Foundation (HON)
International Conference on Harmonization
International Health Terminology Standards Development Organization (IHTSDO)
International Standards Organization (ISO)
Internet Corporation for Assigned Names and Numbers (ICANN)
National Committee for Quality Assurance (NCQA)
National Information Standards Organization (NISO)
National Institute for Standards and Technology (NIST)
National Quality Forum (NQF)
Office of the National Coordinator for Health Information Technology Authorized Testing Body (ONC-ATB)
SNOMED International
Standard development organizations (SDOs)
Workgroup on Electronic Data Interchange (WEDI)
World Wide Web Consortium (W3C)

Statistical Test

A mathematical method designed to help make a quantitative decision about differences between two or more groups of measurements or processes. The intent is to determine whether there is enough evidence (e.g., a large enough difference between the measurements or processes in each group while taking into consideration potential inaccuracies in the measurements) to "reject" a conjecture or hypothesis about the measurement or process. The conjecture is called the null hypothesis (i.e., there is no difference between the two groups).

Analysis of variance (ANOVA)
Area under the curve (AUC)
Bonferroni correction
Case-mix normalization
Chi-square test
Coefficient of variation
Correlation
Correlation coefficient
Cronbach's alpha
Goodness of fit
Kappa value
Kruskal–Wallis one-way analysis of variance
Least squares fitting
Likelihood ratio
Logistic regression
Mann–Whitney test
Mean average precision (MAP)
Mean square error
Measures of concordance
Measures of discordance
Nonparametric test
p-Value
Paired comparison
Parametric test
Receiver operating characteristic (ROC) curve
r-Squared
Stasis statistical test
Statistical Process Control (SPC)
Student's t-test
Survivorship bias
Wilcoxon statistic
z-Score
z-Test

Statistics

The science concerned with the collection, analysis, interpretation, presentation, and organization of data. One of its main functions is to help scientists measure, control, and communicate uncertainty so as to help them learn (i.e., to separate fact from chance) from their data. Statistical methods can be used to help solve a wide variety of scientific, social and business problems.

80/20 rule
A priori probability
Absolute risk
Accuracy
Aggregate
Allocation bias
Anomalous
Artifact
Baseline measurement
Belief network
Categorical data
Causal factor
Centrality
Chance
Clinical subgroup
Clinically relevant population
Cluster
Clustering
Composite estimation
Conditional event
Conditional independence
Conditional probability
Confidence interval
Confidence limits
Contingency table
Cross validation
Cumulative scaling
Curve fitting
Curvilinear
Data interpretation
Data normalization
Decile
Degrees of freedom
Delta
Density coefficients
Dependent variable
Derived parameter

Descriptive variable
Effect size
Error bars
Error function
Estimator (biased, unbiased)
Expected value
External validity
False negative
False negative rate (FNR)
False positive
False positive rate (FPR)
Frequency
Frequency-amplitude domain
Generalizability
Group
Guttman scaling
Hypothesis testing
Independent
Independent variable
Internal validity
Interobserver variation
Interrater reliability
Likert scale
Mean
Median
Metropolitan statistical area (MSA)
Mode
Model
Modeling
Modeling uncertainty
Negative predictive value
Nonsignificant (NS)
Nonsampling error
Nonstationary signals
Normalization
Normalize
Null hypothesis
Null values
Odds
Odds likelihood form
Odds ratio
Odds ratio form
Outcome measure
Outcome variable
Parzen windowing method

Percentile
Polynomial curve fitting
Pool
Pooled data
Population segmentation
Positive predictive value
Posterior probability
Posttest probability
Predictive model
Predictive value
Pretest probability
Prevalence
Prior probability
Probabilistic relationship
Probability
Quartile
Random error
Range
Ratio adjustment
Regression to the mean
Relative risk
Reliability
Reliability estimate
Sampling error
Sampling variance
Scalogram analysis
Scoring
Sensitivity
Sensitivity analysis
Sensitivity calculation
Severity classification
Significance level
Significance testing
Specific
Specificity
Standard deviation
Standard error
Standard error of the mean
Stationary signals
Statistical error
Stochastic
Strata (State Stratification)
Synthetic estimates

Study Design

The process by which experiments, trials of different interventions, or observational studies are designed, developed, and implemented. The goal of a study is to either help the researcher better understand the issue under examination or to assess the safety, efficacy, or mechanism of action of an investigational product, medication, or device. There are many different types of study designs.

Adverse selection
Anonymous reporting
Assignment
Before-after study
Biased selection
Boot-strapping
Case Mix
Case severity
Case-control
Citation analysis
Clinical equipoise
Clinical trial
Cognitive interviewing
Cognitive task analysis
Cognitive walk through
Cohort study
Comorbidity
Comparison-based approach
Conceptual model
Conjoint analysis
Control group
Convenience sample
Cost-effectiveness analysis
Critical experiment
Crucial experiment
Decision facilitation approach
Delphi method
Demonstration study
Descriptive study
Deterministic
Discourse
Distributed research network (DRN)
Double-blind study
Effective sample size
Effectiveness
Efficacious

Efficacy
Emergent property
Ethnographic study
Experiment
Experimental design
Favorable selection
Focus group
Formal systems analysis
Gold standard
Hawthorne effect
Healthcare outcomes
Heuristic evaluation
Hindsight bias
Historical controls
Historically controlled study
Homophily
Human subjects
Hypothesis
In silico
In situ
In vitro
Log analysis
Measurement study
Member checking
Meta-analysis
Modified Delphi method
Monte Carlo simulation
Multistage probability sample
Naturalistic
Number needed to treat
Nyquist frequency
Oral history interview
Orienting issues
Orienting questions
Outcomes
Panel survey
Participatory action research (PAR)
Pattern analysis
Placebo
Placebo effect
Plan, Do, Study, Act (PDSA) cycle
Primary sampling unit (PSU)
Prospective study
Protocol

Purposive sampling
Qualitative data analysis
Qualitative methods
Qualitative model
Quantitative data analysis
Quantitative methods
Random allocation
Randomization
Randomized clinical trial (RCT)
Randomly
Rapid assessment process (RAP)
Reductionist approach
Representativeness
Research protocol
Retrospective chart review
Retrospective study
Sample attrition rate
Sample size
Sample size calculation
Sampling
Screening
Segmentation
Selection bias
Selectivity
Semistructured interviews
Simulation
Site visit
Snowball survey technique
Structured interview
Study population
Study protocol
Subject
Surveillance methods
Survey
Test data set
Think aloud protocol
Time and motion study
Time-motion analysis
Triangulation
Unstructured interview
Usability inspection
Word cloud analysis
Work sampling study

Surgical Procedure

A medical procedure involving an incision with instruments. Such procedures are generally performed to repair damage or arrest disease in a living body. Most surgical procedures are performed under sterile conditions, to reduce the threat of infection, with some type of anesthesia that blocks the patient's pain receptors.

Ablation
Adenoidectomy
Amputation
Angioplasty
Arthroplasty
Atherectomy
Biopsy (Bx)
Biopsy of bronchus
Breast biopsy
Broken bone repair
Caesarean section (C-section)
Cardiac catheterization
Cataract removal
Cholecystectomy (gallbladder removal)
Circumcision
Colonoscopy
Colposcopy
Common bile duct exploration
Coronary artery bypass graft (CABG)
Cryosurgery
Cystoscopy
Debridement of wound, infection, or burn
Decompression peripheral nerve
Diagnostic dilatation and curettage (D&C)
Endoscopic surgery
Endoscopy
Endoscopy of the urinary tract
Esophageal dilatation
Excise
Excision of cervix and uterus
Excision of semilunar cartilage of knee
Femoral hernia repair
Gastroscopy
Hand surgery
Hemilaminectomy
Hysterectomy
Image-guided surgery

Implants
Incision and drainage, skin and subcutaneous tissue (I&D)
Inguinal hernia repair
Joint replacement
Knee cartilage replacement therapy
Laminectomy
Laparoscopy
Laryngoscopy
Laser-assisted in situ keratomileusis (LASIK)
Ligate
Lobotomy
Lumpectomy of breast
Myringotomy (ear tube surgery)
Neovaginoplasty
Partial excision bone
Percutaneous transluminal coronary angioplasty (PTCA)
Radiosurgery
Sigmoidoscopy
Stent procedure
Stereotactic surgery
Suture
Tonsillectomy
Total knee replacement (TKR)
Transurethral removal urinary obstruction
Upper gastrointestinal endoscopy
Ureteral catheterization
Vaginoplasty
Xenotransplantation

System Implementation

The clinical information system implementation process encompasses analyzing requirements, designing new workflows, purchasing hardware and software, installing, configuring, customizing, testing, and training users on both the hardware and software required to make something happen. The word "deployment" is often used as a synonym.

Acceptance testing
Adopt, implement, upgrade (certified EHR Technology)
Analysis phase
Big Bang
Broad and shallow
Build phase
Competency testing
Conformance testing
Data conversion
Data migration
Debriefing
Decommissioning systems
Deployment
Design phase
Document-centric information exchange
Empirical testing
End-user testing
Foundational interoperability
Functional testing
Functionally comparable data models
Historical data
Implementation
Implementation phase
Integrating data
Integration assessment
Integration testing
Interfacing data
Interoperability
Late adopter
Late majority
Legacy system
Luddite
Maintenance phase
Narrow and deep
Nudge
Optimization phase

Phased implementation
Phased installation
Postmortem
Regression testing
Semantic interoperability
Specification phase
Synchronizing content
Syntactic interoperability
System review form
System testing
Systems requirement planning
Technical characteristics
Test patient
Test script
Testing
Testing phase
Train the trainer
Transparency
Trialability
Unit testing
User rights and responsibilities
Zztest

Terminology

The field of study concerned with the systematic development, management, and interrelationships of specific terms and their use to define, label, and describe items, events, actions, and people, for example. These terms can consist of single words, compound words, or multiword expressions that in specific contexts are given specific meanings. Within a specific context or domain, the definition of these terms may deviate from the meanings the same words have in other contexts, domains, or even in everyday language.

Abstraction
Antonym
Canonical form
Child relationship
Clinical modifications
Coding scheme
Component-of relationship
Controlled terminology
Deprecated term
Eponym
Global unique identifiers (GUIDs)
Is-a relationship
Isomorphic data exchange
Kind-of relationship
Language
Lingua franca
Measured-by relationship
Measures relationship
Multiaxial terminology
Nomenclature
Nonsemantic concept identifiers
Not Otherwise Classified (NOC)
Ontology
Parent
Part-of relationship
Polyhierarchy
Polysemy
Postcoordination
Pragmatics
Precoordination
Relationship
Semantic relationship
Semantics
Sibling
Standardize coding and classification

Sublanguage
Synonymy
Syntactic
Syntax
Taxonomy
Terminology authority
Terminology services
Thesaurus
Translation
Treated-by relationship
Treats relationship
Typology
Vocabulary
Work domain ontology (WDO)

Theory

An idea or coherent group of tested propositions, commonly regarded as correct, that are subject to further experimentation before they can be formally accepted as fact, or proven to be true. Theories are often used to provide the basis for an explanation of specific phenomena or the prediction of future phenomena.

Actor–network theory (ANT)
AORTIS (Aggregate, Organize, Reduce, Transform, Interpret, Synthesize) model of clinical summarization
Bayes' theorem
Blackboard architecture
Centering theory
Chaos
Complex adaptive systems (CAS)
Complexity theory
Computability
Data, information, knowledge, wisdom
DeLone and McLean model of information systems
DeMorgen's theorem
Dempster–Shafer theory
Distributional semantics
Empiricism
First law of informatics—do not reuse data
Fundamental theorem of informatics
Gartner Hype Cycle
Grounded theory
Health record banking model
Holism
Intuitionist-pluralist
Just-in-time information model
Logical positivist
Negligence theory
Nyquist theorem
Occam's razor
Paradigm
Principle
Prochaska's Stages of Change
Publish and subscribe model
Roger's diffusion of innovation theory
Shannon's information theory
Sociotechnical model of safe and effective health information technology implementation and use
Systems Engineering Initiative for Patient Safety (SEIPS) model

Technology acceptance model (TAM)
Teleological
Theory of planned behavior
Trellis architecture
TURF (task, user, representation, and function)
Turing test
Unified theory of acceptance and use of technology (UTAUT)
Zipf's law

Unified Medical Language System Vocabulary

The UMLS, or Unified Medical Language System, is one of the crowning achievements of the US National Library of Medicine (NLM). It consists of a set of files and software that brings together many health and biomedical vocabularies and standards to enable interoperability between computer systems. The UMLS has been used to facilitate linking health information, medical terms, drug names, and billing codes to create or enhance applications, such as electronic health records, patient classification tools, clinical dictionaries, and medical language translators.

Alcohol and Other Drug Thesaurus

Alternative Billing Concepts

Anatomical Therapeutic Chemical Classification System

Authorized Osteopathic Thesaurus

Beth Israel Vocabulary

BioCarta online maps of molecular pathways

Biomedical Research Integrated Domain Group Model

Cancer Research Center of Hawaii Nutrition Terminology

Cancer Therapy Evaluation Program—Simple Disease Classification

Canonical Clinical Problem Statement System

Clinical Care Classification

Clinical Classifications Software

Clinical Element Model (CEM)

Clinical Terms Version 3 (CTV3) (Read Codes)

Code on Dental Procedures and Nomenclature

Common Terminology Criteria for Adverse Events

Concept Unique Identifier (CUI)

Consumer Health Vocabulary

COSTAR

COSTART

CRISP Thesaurus

Current Dental Terminology (CDT)

Current Procedural Terminology (CPT)

Definition

Diagnostic and Statistical Manual of Mental Disorders, Fifth Edition (DSM-V)

Diagnostic and Statistical Manual of Mental Disorders, Fourth edition (DSM-IV)

Diagnostic and Statistical Manual of Mental Disorders, Third edition (DSM-III-R)

Diseases Database

FDA National Drug Code Directory

FDA National Drug File

FDB MedKnowledge (formerly NDDF Plus)
Foundational Model of Anatomy Ontology
Gene Ontology
Gold Standard Drug Database
HCPCS Version of Current Dental Terminology (CDT)
Healthcare Common Procedure Coding System (HCPCS)
HL7 Vocabulary
Home Health Care Classification
HUGO Gene Nomenclature
International Classification for Nursing Practice (ICNP)
International Classification of Diseases, 10th Edition, Clinical Modification (ICD-10-CM)
International Classification of Diseases, Ninth Revision, Clinical Modification (ICD-9-CM)
International Classification of Functioning, Disability, and Health (ICF)
International Classification of Functioning, Disability, and Health for Children and Youth
International Classification of Primary Care
International Statistical Classification of Diseases and Related Health Problems
Jackson Laboratories Mouse Terminology
KEGG Pathway Database
Korean Standard Classification of Disease
Library of Congress Subject Headings
Master Drug Data Base
MEDCIN
Medical Dictionary for Regulatory Activities Terminology (MedDRA)
Medical Entities Dictionary
Medical Subject Headings (MeSH)
Medical vocabularies
MedlinePlus Health Topics
Micromedex RED BOOK
Multum MediSource Lexicon
NANDA nursing diagnoses: definitions & classification
National Cancer Institute (NCI) Developmental Therapeutics Program
National Cancer Institute (NCI) Dictionary of Cancer Terms
National Cancer Institute (NCI) Division of Cancer Prevention Program
National Cancer Institute (NCI) SEER ICD Neoplasm Code Mappings
National Cancer Institute (NCI) Thesaurus
National Cancer Institute Nature Pathway Interaction Database
National Center Biomedical Information (NCBI) Taxonomy
National Council for Prescription Drug Programs (NCPDP)

National Drug Codes (NDC)
National Drug File Reference Terminology (NDF-RT)
NeuroNames Brain Hierarchy
North American Nursing Diagnosis Association Taxonomy (NANDA)
Nursing Interventions Classification (NIC)
Nursing Outcomes Classification (NOC)
Omaha system
Online Congenital Multiple Anomaly/Mental Retardation Syndromes
Online Mendelian Inheritance in Man (OMIM)
Patient Care Data Set
Perioperative Nursing Data Set
Pharmacy Practice Activity Classification
Physician Data Query
Physicians' Current Procedural Terminology, Spanish Translation
Read codes
Read thesaurus, American English Equivalents
Read thesaurus, Synthesized Terms
Registry Nomenclature Information System
RxNorm Vocabulary
SNOMED Clinical Terms, Spanish Language Edition
Source of Payment Typology
Specialist lexicon
Standard Product Nomenclature
Systemized nomenclature of medicine (SNOMED)
Systemized nomenclature of medicine clinical terminology (SNOMED-CT)
Systemized Nomenclature of Pathology (SNOP)
Thesaurus of Psychological Index Terms
Traditional Korean Medical Terms
US Centers for Disease Control and Prevention (CDC)
U.S. Food and Drug Administration (FDA)
UltraSTAR
UMDNS: product category thesaurus
UMLS Metathesaurus
Unified Code for Units of Measure (UCUM)
University of Washington Digital Anatomist
USP Model Guidelines
VA National Drug File
Vaccines Administered
Veterans Health Administration National Drug File
Veterinary Extension to SNOMED CT
World Health Organization (WHO) Adverse Reaction Terminology
Zebrafish Model Organism Database

Workflow

A predefined, coordinated, and repeatable pattern of activities facilitated by the systematic organization of physical, human, or information resources into processes that can transform materials, provide services, or process information. It is often depicted as a sequence of operations that one or more agents (i.e., people or computer programs) carry out to accomplish a specific task or set of tasks.

Actors
ADCVAANDIML (Admit, Diagnosis, Condition, Vital signs, Allergies, Activity, Nursing, Diet, IV fluids, Medications, Labs/procedures)
Admission
Advance care planning
Against medical advice (AMA)
Agents
Ambulatory Setting
Appointment
Business Process Modeling Notation (BPMN)
Capacity
Care pathway
Care plan
Care process
Change of shift/report
Clinical event
Clinical feedback
Clinical integration
Clinical pathway
Clinical process
Clinical process model
Clinical scenario
Clone a note
Compromised care process
Computer-based clinical protocol
Consent (informed or patient)
Continuity of care
Continuum of care
Data workflow
Diagnosis (Dx)
Diagnostic hypothesis
Diagnostic process
Direct data entry (DDE)
Direct patient care

Disaster drill
Discharge (DC)
Duplication in, duplication out (DIDO)
Electronic communication
Emergency department/room on divert
Encounter
External hospital transfer
Group visit
Healthcare team
Identical, related, and similar (IRS)
Immediate access
Indirect care
Individual instruction
Information reconciliation
Interdisciplinary care
Internal hospital transfer
Mapping physical locations
Medical record
Medication reconciliation
Messenger
Multidisciplinary care
Multitasking
Nursing care plan
Nursing intervention
Observation
Patient chart
Patient experience
Patient record
Patient triage
Patient-centered medical home (PCMH)
Personal care
Physical artifacts
Point of service
Practice parameter
Precede–proceed
Primary care
Process
Process modeling
Prognosis
Queuing
Register
Report generation
Request for appointment
Rounding

Rounds
Scribe
Secondary care
Shift
Sign
Standard of care
Stat
Structured encounter form
Summary care record
Surveillance
Sweep
Systems analysis
Task
To be determined (TBD)
Transcription
Transitions of care
Treatment plan
Turn around document
Unit dose dispensing
Unit dosing
User acceptance testing (UAT)
Work breakdown structure (WBS)
Work-arounds
Workflow analysis
Workflow elements model (WEM)
Workflow model
Working diagnosis

INDEX

H

H&H—Hemoglobin and Hematocrit, 136

H&P—History and Physical, 26

Habituation, 138

HAC—Hospital Acquired Condition, 115

Hacker, 126

Hadoop, 40

Haemophilus influenza type B vaccine (Hib), 121

Haemophilus influenza, 69

HAI—Healthcare Acquired Infection, 143

Hair, 5

Half-life, 13

Halting problem, 108

Hammond, W. Edward, 132

Hand surgery, 157

Handheld device, 45

Handicapped, 22

Hand-off, 72

Handwriting recognition, 90

Hanmer, Lyn, 132

Hannah, Kathryn, 132

Hannan, Terry, 132

Haptic feedback, 88

Hard disk, 45

Hard of Hearing (HOH), 23

Hard stop, 18

Hardware, 45, 48

Harm, 129

Hash function, 37

Hash table, 51

Hash value, 108

Hashed array tree, 63

Hashing, 51

Hasman, Arie, 132

Haux, Reinhold, 132

Hawthorne Effect, 155

Haynes, R. Brian, 132

Hazard Log, 129

Hazard, 129

HbA1C—Hemoglobin A1C or Glycohemoglobin, 135

Hb—Hemoglobin, 135

HCA—Hospital Corporation of America, 59

HCAHPS—Hospital Consumer Assessment Healthcare Providers and Systems, 94

HCBS—Home and Community-Based Services, 128

HCFA—Health Care Financing Administration, 81

HCI—Human computer interaction, 76

HCO—Health care organization, 128

HCPCS Version of Current Dental Terminology (CDT), 166

HCPCS—Healthcare Common Procedure Coding System, 166

HCTZ—Hydrochlorothiazide, 121

HCUP QIs—Healthcare Cost and Utilization Project Quality Indicators, 143

HCV—Hepatitis C Virus, 69

HDL-C—High-Density Lipoprotein cholesterol test, 136

HDL—High Density Lipoprotein, 136

Head, Eyes, Ears, Nose, (Mouth), and Throat (HEENT), 86

Headache, 23

Header of message, 35

Health Care Financing Administration (HCFA), 81

Health care organization (HCO), 128

Health Care Paraprofessional, 30

Health care team, 169

Health data broker, 30

Health data custodian, 30

Health Education, 86

Health evaluation (Apache—II) scoring system, 135

Health Evaluation through Logical Programming (HELP), 40

Health Facilities, 118

Health Informatician, 30

Health Informaticist, 30

Health informatics Service Architecture (HISA), 43

Health Informatics Standards Board (HISB), 148

Health Informatics, 76

Health information access layer (HIAL), 43

Health Information Exchange (HIE), 40

Health Information Management (HIM), 76

Health Information Management and Systems Society (HIMSS), 141

Health Information Technology (HIT), 76

Health Information Technology for Economic and Clinical Health (HITECH) Act, 100

Health Insurance Flexibility and Accountability (HIFA), 100

Health Insurance Portability and Accountability Act (HIPAA), 1996, 100

Health Insurance Purchasing Cooperative (HIPC), 83

Health Insurance, 83–84

Health IT System, 130

Health Level 7 (HL7) analyst, 132

Health Level Seven (HL-7 or HL7), 148

Health Maintenance Organization (HMO), 128

Health Manpower Shortage Area (HMSA), 80

Health on the Net Foundation (HON), 148

Health Personnel, 30

Health Plan Employer Data and Information Set (HEDIS), 143

Health Plan, 83

Health Planning, 105

Health Policy, 76

Health Promotion, 86

Health record Banking model, 163

Health Resources and Services Administration (HRSA), 81

Health Risk Assessment (HRA), 72

Health Risk Factors, 84

Printed in the United States
By Bookmasters